Growth Hormone - Impact and Insights in Human Beings

Edited by Mario Bernardo-Filho, Danúbia da Cunha de Sá-Caputo, Técia Maria de Oliveira Maranhão and Redha Taiar

Published in London, United Kingdom

Growth Hormone - Impact and Insights in Human Beings
http://dx.doi.org/10.5772/intechopen.102210
Edited by Mario Bernardo-Filho, Danúbia da Cunha de Sá-Caputo, Técia Maria de Oliveira Maranhão and Redha Taiar

Contributors
Jesús Devesa, Pablo Devesa, Rivada M. Kurabekova, Olga P. Shevchenko, Sergey V. Gautier, Olga M. Tsirulnikova, Irina E. Pashkova, Mithal Dhbea, Abdilya Alabdaly, Mario Bernardo-Filho, Danúbia da Cunha de Sá-Caputo, Redha Taiar, Tecia Maria de Oliveira Maranhao

First published in London, United Kingdom, 2023 by IntechOpen
IntechOpen is the global imprint of INTECHOPEN LIMITED, registered in England and Wales, registration number: 11086078, 5 Princes Gate Court, London, SW7 2QJ, United Kingdom

British Library Cataloguing-in-Publication Data
A catalogue record for this book is available from the British Library

Additional hard and PDF copies can be obtained from orders@intechopen.com

Growth Hormone - Impact and Insights in Human Beings
Edited by Mario Bernardo-Filho, Danúbia da Cunha de Sá-Caputo, Técia Maria de Oliveira Maranhão and Redha Taiar
p. cm.
Print ISBN 978-1-83768-268-3
Online ISBN 978-1-83768-269-0
eBook (PDF) ISBN 978-1-83768-270-6

Meet the editors

Professor Mario Bernardo-Filho heads the Laboratory of Mechanical Vibrations and Integrative Practices (LAVIMPI), Roberto Alcantara Gomes Biology Institute, Department of Biophysics and Biometry, Piquet Carneiro University Polyclinic, State University of Rio de Janeiro (UERJ). He obtained his Ph.D. in 1988 and has been a full professor at UERJ since 1974. He supervises MSc and Ph.D. theses and has been a speaker at various national and international conferences. He is Editor in Chief of the *Brazilian Journal of Mechanical Vibrations in Biosciences* and a reviewer for several scientific journals indexed in PubMed and Scopus. He is interested in subjects related to exercises/physical activities and their biological effects in healthy and unhealthy individuals of different ages.

Redha Taiar, Ph.D., is a professor at the University of Reims Champaign, France. His research focuses on engineering for medicine and high-level sport. He has worked as an engineer for various industries including Arena for high-level sport and Sidas, Medicapteur for medical development. His most recent work in the industry involved the design and testing of Notrax anti-fatigue mats. In sport his most recent work focuses on the development of swimsuits for triathletes and swimmers at the Brazil Olympic Games (2016) and Paris 2024, and ski-suit fabrics for the Winter Olympics at Sotchi in 2014. He is a specialist in the biomechanics of health disease and rehabilitation.

Prof. Técia Maranhão is a retired professor in the Obstetrics and Gynecology Department of the Science Health Center of the Federal University of the Rio Grande do Norte (UFRN). She obtained her Ph.D. in Obstetrics and Gynecology from the Faculty of Medicine of Ribeirão Preto, University of São Paulo in 1982. She is a member of the National Commission of Endocrine Gynecology of the Brazilian Federation of Obstetrics and Gynecology and is on the Editorial Board of the *RBGO*. She lectures at various national and international conferences and is on the teaching staff of the Professional Master's Degree in Health Education. She works in the area of reproductive health and endocrine gynecology, and is currently researching polycystic ovary syndrome, insulin resistance and metabolic syndrome, and changes in female physical performance in hyperandrogenism.

Prof. Danúbia da Cunha de Sá-Caputo is a Research Project Coordinator at the Laboratory of Mechanical Vibrations and Integrative Practices (LAVIMPI) at the State University of Rio de Janeiro UERJ, Brazil. She obtained her MSc in 2015 and PhD in 2016. She supervises MSc and Ph.D. students and is a speaker at various national and international conferences. She is a reviewer of several scientific journals indexed in PubMed and Scopus. Her research interests relate to exercises/physical activities and their biological effects on healthy and unhealthy individuals of different ages.

Contents

Preface

Growth hormone (GH) is synthesized and secreted daily by anterior pituitary cells called somatotrophs. GH has a role to play in various physiological processes. It is fundamental for the normal physical growth of children and, irrespective of the age of the individual, it is necessary for the proper functionality of the metabolic responses. GH stimulates the synthesis of proteins and increases the breakdown of fat mass to generate the energy necessary for tissue growth. It also counterpoints the action of insulin. While it appears to act on different tissues, part of its effect is mediated by the release of insulin-like growth factor-1 (IGF-1). IGF-1 can mimic the action of insulin, but its primary action is also the stimulation of growth.

Some individuals are affected by abnormalities in GH secretion, which can be either deficient or excessive. Production of GH in individuals can be stimulated by different types of physical exercises, and GH deficiency can be treated with direct administration of GH into individuals. GH deficiency is also a cause of short stature and dwarfism in some individuals. GH deficiency that persists into adulthood is associated with various undesirable clinical conditions, such as fatigue, decreased energy, decreased muscle strength, decreased muscle mass, thin and dry skin, increased adipose tissue, which favors obesity, and decreased bone density. Excess GH production may be due to a benign tumor of the somatotroph cells of the pituitary gland, as well as tumor of the lung or of the pancreatic islets. Excess of GH may lead to extreme height (gigantism) and features of acromegaly.

This book, which is in three sections, introduces the reader to the biochemistry of growth hormone and its relevance to human beings. Section 1 highlights the relevance of the relationship between GH, obesity, and physical exercise. Section 2 consists of two chapters: "Growth Hormone Gene Family and Its Evolution" and "Pituitary Growth Hormone Secretion and Cell Growth Hormone Production: Regulation of Their Secretion and Their Signaling Pathways". Section 3 discusses the types of interventions in three chapters: "Growth Hormone and Insulin-like Growth Factor-1 in Children with Cholestatic Diseases and Pediatric Liver Transplantation", "Carbohydrate Metabolism in Growth Hormone Therapy for Children" and "Obesity: The Relationship between Growth Hormone and Exercises".

The editors would like to thank Author Service Manager Maja Bozicevic at IntechOpen for her amazing work, and for her patience and assistance throughout the preparation of this book.

Mario Bernardo-Filho and Danúbia da Cunha de Sá-Caputo
Laboratory of Mechanical Vibrations and Integrative Practices,
Roberto Alcantara Gomes Biology Institute,
Piquet Carneiro University Polyclinic,
State University of Rio de Janeiro,
Rio de Janeiro, Brazil

Redha Taiar
Materials and Mechanical Engineering Research Group,
University of Reims,
Reims, France

Técia Maria de Oliveira Maranhão
Department of Obstetrics and Gynecology,
Center of Health Sciences,
Federal University of Rio Grande do Norte,
Natal, Brazil

Section 1

Introduction

Chapter 1

Introductory Chapter: Growth Hormone – Obesity and Physical Exercise

Mario Bernardo-Filho, Redha Taiar, Técia Maria de Oliveira Maranhão and Danúbia da Cunha de Sá-Caputo

1. Introduction

Obesity is one of the most prevalent metabolic diseases in the recent decades leading to negative effects on health. Obesity has shown a significant global increase over the past 50 years and this fact would be not desired. The World Health Organization (WHO) defines obesity as when an individual has a body mass index (BMI) greater than 30 kg/m^2. The epidemic growth of obesity represents a very serious public health problem, mainly due to its important association with the development of serious diseases that can lead to death, such as type 2 diabetes, cardiovascular diseases, arterial hypertension, metabolic syndrome, chronic kidney disease, hyperlipidemia, some types of cancer, asthma, sleep disorders, liver dysfunction, kidney disease, joint disease, depression, and infertility [1, 2].

Obesity is a universal phenomenon that occurs globally, except for parts of Asia and sub-Saharan Africa, as well as a few other countries, such as, Sri Lanka, Indonesia, Sudan, Singapore, and Djibouti [1].

It is known that several complex mechanisms can lead to obesity. The main cause would be the occurrence of an excess of energy stock when compared with the energy needed by the body. Excess energy, stored in fat cells, which enlarge, alters metabolic and hormonal factors responsible for obesity. Studies reveal that the food source and quality in the diet seem to influence the body mass control more than the amount ingested. Moreover, growing evidence suggests that obesity is a disorder of the energy homeostasis system. Furthermore, the lack of balance between food intake and physical activity, between genetic and epigenetic, environmental, and microenvironmental factors, the availability of health care services, and socioeconomic status can lead to obesity [3–5].

It is already well established that the body location of adipose tissue, mainly in the abdominal region, the so-called central obesity (CO), represents an important risk factor for metabolic dysfunction. It is significantly correlated with cardiovascular disease. CO can be classified into subcutaneous abdominal adipose tissue (SAT) and visceral adipose tissue (VAT). It is suggested that VAT is a metabolic organ that regulates fat mass, blood glucose, and nutrient homeostasis. Based on this knowledge, it would be really important to identify clinical means capable of improving the therapeutic approach to obesity [1, 5–8].

Diet modifications and physical exercise are always included as a priority in treatments for obesity. Physical exercise would promote changes in lipolytic hormones, such as growth hormone (GH) and significant reductions in VAT. VAT, regardless of total BMI, is strongly related to cardiovascular risk [9–12].

The relevance of hormones in controlling body metabolism has been known for a long time. Among the most kinetically studied hormones is GH. Some aspects of GH physiology and its relationship with exercise, however, still remain controversial and unknown, including the knowledge of elements that regulate its synthesis, mechanisms of action and effects on protein and lipid metabolism [13, 14].

GH is secreted by the anterior pituitary in a pustile form, primarily regulated by the hypothalamic GH-releasing hormone (GHRH) and by somatostatin, neuropeptides that respectively stimulate and inhibit the secretion of this hormone. It is known that GH affects several body systems, regulating the secretion of insulin-like growth factor-1 (IGF-1), which in turn regulates its secretion. GH is a potent anabolic hormone with an important role in lipid metabolism in various body sites such as the liver, muscle skeletal muscle, and adipose tissue [15–17].

In times of fasting or stress, GH promotes the use of lipids as the first source of energy in order to preserve carbohydrate and protein stores. Several other physiological and pathological conditions interfere with GH secretion, age, gender, pubertal stage, sleep status, nutritional status, body composition and temperature, fitness, gonadal steroids, insulin, and IGF-1. In the physiology of GH and its relationship with the aforementioned factors, much remains to be clarified, and this will be highly relevant. It is known that abdominal visceral fat increases in obese individuals, which is why there is a drastic reduction in the GH response in obese individuals, both in stimulated secretion and in normal release [6, 14, 18, 19].

Since physical exercise is considered a potent stimulator of GH synthesis [20], factors such as exercise intensity, volume, and frequency may influence in GH concentration. However, studies have shown that in obese individuals, the GH elevation in response to physical exercise is reduced and that with body mass loss the response returns to that found in individuals with normal BMI [21, 22].

Regular exercise is desirable for health and studies have shown that different kinds of exercise can aid in the reduction of the body fat as well as protect the individuals against cardiovascular diseases associated with obesity [23–25].

2. Conclusion

It is known that GH acts directly through its receptor and indirectly stimulates the production of IGF-1. However, the mechanism through which physical exercise can be used to increase GH secretion is not completely understood. In this sense, this book aims to integrate scientific information about the relevance of exercises and/or GH to the management of individuals with obesity. Moreover, it is expected that this book might bring new insights into the role of growth hormone in the obesity.

Author details

Mario Bernardo-Filho[1*], Redha Taiar[2], Técia Maria de Oliveira Maranhão[3] and Danúbia da Cunha de Sá-Caputo[1]

1 Laboratory of Mechanical Vibrations and Integrative Practices, Roberto Alcantara Gomes Biology Institute, Piquet Carneiro University Polyclinic, State University of Rio de Janeiro, Rio de Janeiro, Brazil

2 Materials and Mechanical Engineering Research Group, University of Reims, Reims, France

3 Department of Obstetrics and Gynecology, Center of Health Sciences, Federal University of Rio Grande do Norte, Natal, Brazil

*Address all correspondence to: bernardofilhom@gmail.com

References

[1] Lin X, Li H. Obesity: Epidemiology, pathophysiology, and therapeutics. Frontiers in Endocrinology (Lausanne). 2021;**12**:706978. DOI: 10.3389/fendo.2021.706978

[2] Hruby A, Hu FB. The epidemiology of obesity: A big picture. PharmacoEconomics. 2015;**33**(7):673-689. DOI: 10.1007/s40273-014-0243-x

[3] Schwartz MW, Seeley RJ, Zeltser LM, Drewnowski A, Ravussin E, Redman LM, et al. Obesity pathogenesis: An endocrine society scientific statement. Endocrine Reviews. 2017;**38**(4):267-296. DOI: 10.1210/er.2017-00111

[4] Moehlecke M, Canani LH, Silva LO, Trindade MR, Friedman R, Leitão CB. Determinants of body weight regulation in humans. Archives of Endocrinology and Metabolism. 2016;**60**(2):152-162. DOI: 10.1590/2359-3997000000129

[5] Theilade S, Christensen MB, Vilsbøll T, Knop FK. An overview of obesity mechanisms in humans: Endocrine regulation of food intake, eating behaviour and common determinants of body weight. Diabetes, Obesity and Metabolism. 2021;**23**(Suppl. 1):17-35. DOI: 10.1111/dom.14270

[6] Swinburn BA, Sacks G, Hall KD, McPherson K, Finegood DT, Moodie ML, et al. The global obesity pandemic: Shaped by global drivers and local environments. Lancet. 2011;**378**(9793):804-814. DOI: 10.1016/S0140-6736(11)60813-1

[7] Sabag A, Chang D, Johnson NA. Growth hormone as a potential mediator of aerobic exercise-induced reductions in visceral adipose tissue. Frontiers in Physiology. 2021;**12**:623570. DOI: 10.3389/fphys.2021.623570

[8] Sacks FM, Bray GA, Carey VJ, Smith SR, Ryan DH, Anton SD, et al. Comparison of weight-loss diets with different compositions of fat, protein, and carbohydrates. New England Journal of Medicine. 2009;**360**(9):859-873. DOI: 10.1056/NEJMoa0804748

[9] Berryman DE, List EO. Growth hormone's effect on adipose tissue: Quality versus quantity. International Journal of Molecular Sciences. 2017;**18**:1621. DOI: 10.3390/ijms18081621

[10] Fox CS, Massaro JM, Hoffmann U, Pou KM, Maurovich Horvat P, Liu CY, et al. Abdominal visceral and subcutaneous adipose tissue compartments: Association with metabolic risk factors in the Framingham heart study. Circulation. 2007;**116**:39-48. DOI: 10.1161/CIRCULATIONAHA.106.675355

[11] Pak K, Lee SH, Lee JG, Seok JW, Kim IJ. Comparison of visceral fat measures with cardiometabolic risk factors in healthy adults. PLoS One. 2016;**11**:e0153031. DOI: 10.1371/journal.pone.0153031

[12] Saxton SN, Clark BJ, Withers SB, Eringa EC, Heagerty AM. Mechanistic links between obesity, diabetes, and blood pressure: Role of perivascular adipose tissue. Physiology Reviews. 2019;**99**(4):1701-1763. DOI: 10.1152/physrev.00034.2018

[13] Stanley TL, Grinspoon SK. Effects of growth hormone-releasing hormone on visceral fat, metabolic, and cardiovascular indices in human studies. Growth Hormones IGF Research. 2015;**25**:59-65. DOI: 10.1016/j.ghir.2014.12.005

[14] Lewitt MS. The role of the growth hormone/insulin like growth factor system in visceral adiposity. Biochemistry Insights. 2017;**10**:1178626417703995-1178626417703995. DOI: 10.1177/1178626417703995

[15] Vijayakumar A, Yakar S, Leroith D. The intricate role of growth hormone in metabolism. Frontiers in Endocrinolology. (Lausanne). 2011;**2**:32. DOI: 10.3389/fendo.2011.00032

[16] Al-Samerria S, Radovick S. The role of insulin-like growth factor-1 (IGF-1) in the control of neuroendocrine regulation of growth. Cell. 2021;**10**(10):2664. DOI: 10.3390/cells10102664

[17] Dehkhoda F, CMM L, Medina J, Brooks AJ. The growth hormone receptor: Mechanism of receptor activation, cell signaling, and physiological aspects. Frontiers in Endocrinology (Lausanne). 2018;**9**:35. DOI: 10.3389/fendo.2018.00035

[18] Al Kibria GM. Prevalence and factors affecting underweight, overweight and obesity using Asian and World Health Organization Cutoffs among adults in Nepal: Analysis of the demographic and health survey 2016. Obesity Research & Clinical Practice. 2019;**13**(2):129-136. DOI: 10.1016/j.orcp.2019.01.006

[19] Frühbeck G, Toplak H, Woodward E, Yumuk V, Maislos M, Oppert JM. Obesity: The gateway to ill health–an EASO position statement on a rising public health, clinical, and scientific challenge in Europe. Obesity Facts. 2013;**6**:117-120. DOI: 10.1159/000350627

[20] Gomes MR, Pires I, Castro IA, Tirapegui J. Effect of moderate physical exercise on plasma and tissue levels of insulin-like growth factor-1 (IGF-1) in adult rats. Nutrition Research. 2004;**24**:555-564

[21] Heiston EM, Eichner NZM, Gilbertson NM, Gaitán JM, Kranz S, Weltman A, et al. Two weeks of exercise training intensity on appetite regulation in obese adults with prediabetes. Journal of Applied Physiology (1985). 2019;**126**(3):746-754. DOI: 10.1152/japplphysiol.00655.2018

[22] Lee BS, Kim KA, Kim JK, Nho H. Augmented hemodynamic responses in obese young men during dynamic exercise: Role of the muscle Metaboreflex. International Journal of Environmental Research and Public Health. 2020;**17**(19):7321. DOI: 10.3390/ijerph17197321

[23] Son WM, Sung KD, Bharath LP, Choi KJ, Park SY. Combined exercise training reduces blood pressure, arterial stiffness, and insulin resistance in obese prehypertensive adolescent girls. Clinical and Experimental Hypertension. 2017;**39**:546-552

[24] Wannamethee SG, Shaper AG, Walker M. Physical activity and mortality in older men with diagnosed coronary heart disease. Circulation. 2000;**102**:1358-1363

[25] Warburton DE, Glendhill N, Quinney A. The effects of changes in musculoskeletal fitness on health. Canadian Journal of Applied Physiology. 2001;**26**:161-216

Section 2

Biochemistry and Physiology

Chapter 2

Growth Hormone Gene Family and Its Evolution

Jesús Devesa and Pablo Devesa

Abstract

In this review, we will analyze the family of growth hormone (GH) genes, the territories where they are produced, the proteolytic generation of GH isoforms, both at the pituitary and tissue levels, the biological activity of these molecular forms, and we will describe the new variant GH-V2 and its effects biological. Finally, we will analyze the evolution of the hormone from its starting point with a common gene with PRL to its actions in the most evolved organisms as a true prohormone.

Keywords: GH gene, GH-N, GH-V, GH-V2, evolution of GH

1. Introduction

Just one century ago (1921), it was reported that the administration of bovine anterior pituitary gland extracts to rats induced greater growth in these thus treated animals [1]. This fact led to the assumption that a pituitary factor had to be responsible of the longitudinal growth of the organism, something proven some years later when it was shown that human dwarfs could grow when treated with human pituitary extracts. This pioneer treatment was introduced by Dr. Maurice Raben at Tufts New England Medical Center in 1956 [2]. One of the first dwarfs treated in this way, who reached normal height, was in good health 62 years later [3]. However, the discovery of the human growth hormone (GH) took place a year earlier, 1955, by Choh Hao Li, who worked at the University of California, within his studies on the isolation of pituitary hormones. Li himself developed a method for isolating GH from human cadavers and purifying the hormone so that it could be administered to children with GH-deficiency dwarfism. At the program's peak in 1973, 82,500 pituitary glands had been collected for treatment of about 3,000 children. Too many pituitary glands removed from human cadavers for so few GH-deficient children treated, so that the program declined. Furthermore, the extraction techniques could not avoid the fact that some of the pituitaries were contaminated with nerve tissue from the posterior pituitary lobe. This led to the appearance of several cases of iatrogenic Creutzfeldt-Jacob disease in patients who had received pituitary extracts in which pathogenic prions were present in the posterior pituitary lobe, something *a priori* undetectable. In 1962, the Li group made the first approach to the amino acid composition of human pituitary GH [4], a very important finding that allowed him to determine the structure of human GH and synthesize it in small quantities in 1971. This was a very important step forward as it paved the way for the production of the hormone in genetically engineered bacteria (*Escherichia Coli*), greatly expanding the offer. The product was approved in the United States for sale in October 1985, and a new world began for GH-deficient children, because the advances in the field of genetically engineering permitted to produce practically unlimited quantities of the hormone, pure and safe, by DNA recombinant technology. The hormone was first produced in prokaryotes (*Escherichia Coli*) and later in eukaryotes (*murine fibroblasts*). Apparently, the hormones produced in these different type of cells were the same in terms of their effectiveness in producing further growth in GH-deficient children. However, we suspected that there must be some differences depending on the type of cells used for production of GH. In fact, similar doses expressed in international units (IU), corresponded to different weight of the theoretically identical product. To analyze the reason for these differences, we carried out an electrophoretic study of the different GHs existing on the market, and we saw that the hormone produced in murine fibroblasts contained not only the main GH variant GH 22 kDa but also the minor variant 20 kDa. This is logical, since during the expression of human GH, an alternative splicing occurs in 10% of the primary transcript, giving rise to the 20 kDa GH, but this alternative splicing cannot take place in prokaryotes. Initially, these differences were not considered important in terms of effectiveness at the growth level; and it took many years until a utility or any physiological action could be found for this GH 20 kDa. However, different studies carried out in the last years reveal that this isoform *in vivo* possesses very important properties. For example, it has been seen that GH 20 kDa has several GH-like activities in male high-fat-fed rats, lacking, unlike GH, diabetogenic and lactogenic effects, and failing to increase plasma IGF-I or body length. Instead, treating mice with GH gene deletion (−/−) leads to clear increases in plasma IGF-I, femur length, body length, body weight, and lean body mass and decreased fat mass. These effects are similar to those observed in mice receiving GH treatment. That is, GH 20 kDa can stimulate IGF-I and longitudinal body growth in GH-deficient mice in a similar way to GH 22 kDa, but unlike this main form, the GH isoform promotes growth without inhibiting the actions of insulin and without promoting (*in vitro*) growth of cancers presenting prolactin receptor (PRLR). The authors conclude that this GH isoform may improve current GH therapies especially in patients at risk for metabolic syndrome or PRLR-positive cancers [5]. Another recent study indicates that GH 20 kDa can internalize into the cytoplasm, as GH 22 kDa does, but its functions appear to be different from those exerted by the main form [6]. From these and other studies, it is expected that in the coming years, the possible beneficial effects of GH 20 kDa on the human body will be better known.

From 1985, when treatments with recombinant GH began, the spectrum of therapeutic applications of this hormone, initially restricted to children with proven GH-deficiency, increased significantly, especially in the pediatric population [7]. This is the case in children suffering from idiopathic short stature, or children with growth retardation caused by chronic renal failure, Turner syndrome, *SHOX* gene deficiency, Noonan syndrome, but also in adults with GH deficiency as long as they have any other hormonal pituitary deficiency (with the sole exception of Prolactin), and in patients with acquired immunodeficiency syndrome (AIDSS) wasting [8, 9]. This led to GH being considered a hormone that, in addition to its metabolic properties (hyperglycemic by counteracting the effects of insulin, lipolytic and protein anabolic), was considered as the growth hormone. Consequently, the study of its effects was practically restricted to pediatric doctors, whose main interest in GH was to prescribe it for growth in children with short stature.

1.1 The new concepts about GH

At that time, relatively few researchers tried to investigate whether the hormone might could have other actions in the body far beyond those described [10]. Perhaps this situation changed when a previous study was analyzed in which it had been discovered that Insulin and Epidermal Growth Factor (EGF) could be internalized in living cells [11], contrary to the classical concept that protein hormones only carry out their biological actions after interacting with specific membrane receptors, that is, without entering cells, unlike what happens with steroid or thyroid hormones whose receptors are located in the cytoplasm or cell nucleus. That finding led to the Waters group, in Queensland, to investigate whether the same could happen with GH [12–14]. They demonstrated that this hormone, after interacting with its membrane receptor (GHR), is also internalized along with its GHR through the endosomal pathway. This mechanism allows the translocation of GH and GHR to the nucleus of the cell where they induce the transcription of many different genes. This was a key discovery, as it allegedly implied that GH could perform many diverse, tissue-specific functions, far beyond its classical known effects. Therefore, the detection of GHR in the nucleus of a cell indicates that there has been a previous interaction GH-GHR at the cell membrane. Another important finding that reinforced the idea that GH exerted many different effects in the organism was the demonstration by our group that internalized GH also undergoes specific tissue proteolytic processing, influenced by sex and age (in mice), which gives rise to different molecular forms, whose actions are unknown still [15]. **Figure 1** schematizes these concepts.

According to these concepts, it is logical to understand that GH is a hormone whose actions can affect practically the entire organism, instead of being a mere metabolic hormone and the hormone for growth. Throughout this chapter, we will

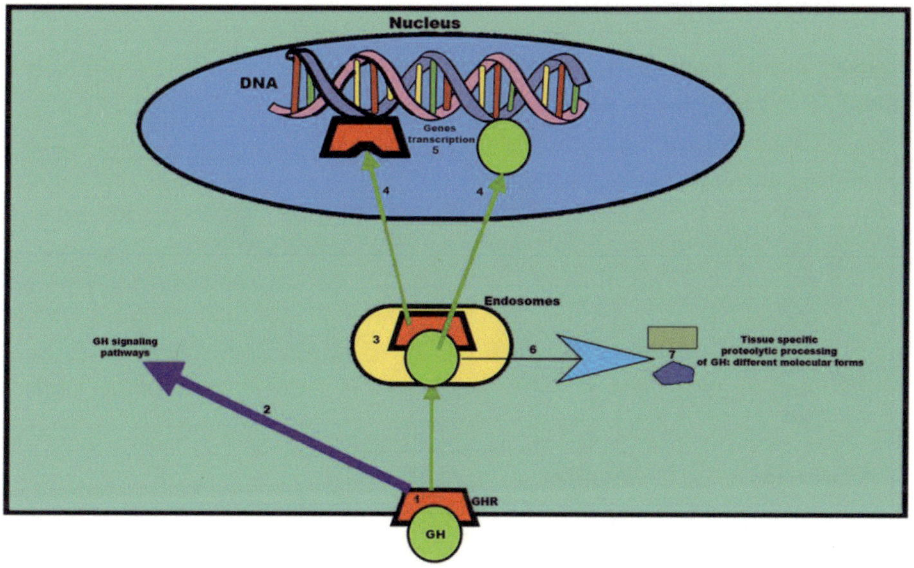

Figure 1.
Interaction GH-GHR and translocation to the nucleus of the cell. (1) Once GH interacts with its membrane receptor GHR, it is activated (red color) and induces a cascade of phosphorylation (2) that represents GH signaling pathways. (3) GH and GHR are internalized through the endosomal pathway. In endosomes, parts of GH and GHR are proteolytically degraded, but another part translocates to the nucleus (4), where they induce gene transcription (5). In endosomes, GH gives rise (6) to different tissue-specific molecular forms (7), whose actions are still unknown.

analyze the role of the genes that make up the GH family, as well as how this hormone is now considered a prohormone given its pleiotropic actions.

2. Family of GH genes

2.1 The GH gene locus

The locus of the human genome where the genes belonging to the GH family are found is located on the long arm of chromosome 17, in the region q22–24 [16–18] (**Figure 2**), spanning 48 kb of DNA in the order: 5′-(hGH-1/-hCS-5/hCS-1/hGH-2/hCS-2)-3′ [19, 20]. These genes encode normal human GH (GH-N) and its variant (GH-V) as well as the group of chorionic genes: hCS-L, hCS-A, and hCS-B. These are genes highly conserved in evolution, being the homology among them 5%. The hGH/hCS gene locus seems that it has been evolved by duplication mechanisms [20]. They come from a common ancestral gene with that of Prolactin (PRL) from which they diverged about 300–400 million years ago, although the gene coding for PRL is found on chromosome 6. All these genes have five exons separated by four introns [20] (**Figure 2**). Among them, GH-1 (now known as GH-N) is expressed in the pituitary gland and peripheral tissues, while the other four seem to be expressed only in the placenta, although N-glycosylated GH-related peptides have been found in human pituitary extracts, suggesting that the hGH-2 gene (now known as GH-variant [GH-V), or other unknown GH-related genes, could be expressed too at the pituitary level, and perhaps in other tissues, since the hGH-N gene lacks the consensus sequence for N-glycosylation observed in some nonplacental GH-related products [21, 22].

2.2 Molecular GH heterogeneity

Figure 2 is only a schematic description of the hGH gene family, but pituitary and tissue GH heterogeneity is really high. For example, in the human pituitary,

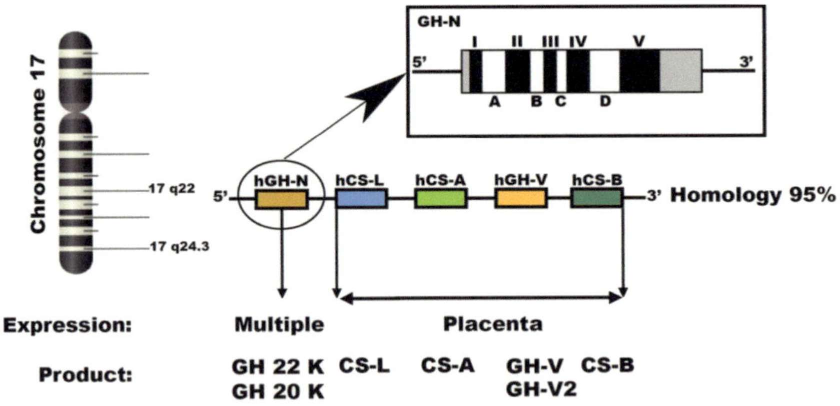

Figure 2.
hGH/hCS gene locus localization, expression of the genes of this family and products resulting from it. As noted above, the hGH/hCS gene family locus is located on the long arm of chromosome 17 (q22–q24). The figure also represents the 5′—3′ arrangement of the genes that make up this family, of which GH-N is outlined in the upper rectangle (indicated by a black arrow). In this gene, it can be seen the organization in exons (I–V) separated by introns (A–D). Note that, although the GH-N gene is expressed practically throughout the body (pituitary and multiple tissues and organs), the other four genes seem to be expressed only in the placenta, although as we will see, this is not exactly the case, at least for the GH-V gene.

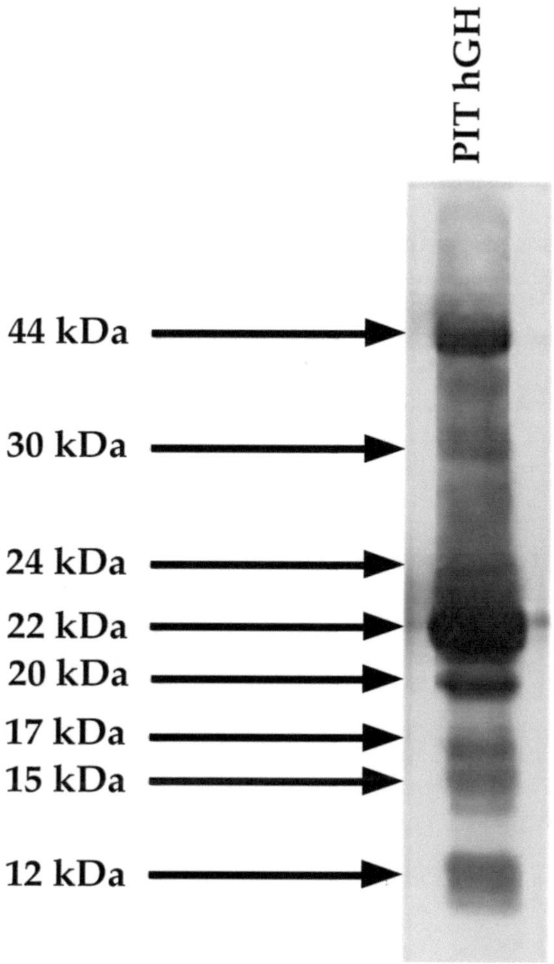

Figure 3.
Heterogeneity of human pituitary (PIT) GH. Western blot showing the high heterogeneity of GH-N obtained from a human pituitary extract. Note the dimer 44 kDa formed by GH 22 kDa, as well as the number of GH-N variants whose biological significance is unknown. The 5 kDa GH form is not shown in this figure.

many isoforms of GH-N can be found. Apart from the main GH products, 22 kDa and the 20 kDa form of GH, the latter resulting from the alternative splicing of the mRNA of the GH-N gene, there are several posttranslationally modified forms of GH (N-acylated, deamidated, and O-glycosylated forms of monomeric GH), as well as non-covalent and disulfide-linked oligomers up to pentameric GH [23, 24]. The high heterogeneity of pituitary GH can be seen in **Figure 3**.

Furthermore, in human pituitary extracts, some significant amounts of GH variants of lower molecular weight (17 kDa and 5 kDa) can be found. These originate from the selective cleavage of the bond between amino acids 43 and 44, leading to the production of fragments 1–43 and 44–191 in the somatotrophic cells themselves and have also been found in human plasma and tissues and in some different species [15, 25–30]. Since these forms can be originated by specific cleavage of the main 22 kDa GH form, it is likely that they appear in tissues as a result of posttranslational modification of the hormone. Interestingly, while human GH 22 kDa has both insulin-like and diabetogenic effects, these proteolytically generated fragments show opposite effects: the short 5 kDa GH form potentiates the effects of the insulin, while the 17 kDa form shows diabetogenic activity [28–30]. Since these two isoforms can be generated under acidic conditions, it has been postulated that they may play a significant physiological role because the hormone in cells is exposed to an acidic environment [25]. Similar conclusions have been drawn regarding the effects of these 17 and 5 kDa GH variants at the end of the last century. According to their data [29–31], these GH fragments, obtained by recombinant DNA technology, have potent in vivo effects on glucose homeostasis in rodents, but cannot stimulate body growth.

The great heterogeneity of forms produced from GH-N is due to the fact that in the main molecule, there are a series of points susceptible of hydrolysis by proteases, which would be another point in favor of considering GH as a prohormone, which undergoes selective proteolysis, both in the pituitary and in specific tissues, giving rise to peptides with specific biological activities. For example, there are two fragments of GH, GH 177–191 and GH 95–133, of which the first shows a lipolytic activity identical to that of the 22 kDa GH-N form, while the second is a potent mitogen. For its part, the 24 kDa GH-N form caused by retention of signal peptide, like the 12 kDa form originating from the previous one, shows greater somatogenic and lactogenic potency than the 22 kDa GH-N itself.

The alternative splicing of the major pituitary GH-N form 22 kDa occurs in approximately 10% of the transcripts in eukaryotes (**Figure 4**).

As a consequence of the cleavage of part of intron III in the 20 kDa GH-N form, the amino acids located from position 32 to 46 have disappeared (**Figure 5**).

The pituitary expression of these two proteins (GH-N 22 kDa and 20 kDa) reflects the alternative utilization of a major (B) and a minor (B′) splice acceptor site in exon 3 of the hGH-N transcript, although it has been postulated that, in addition to the importance of sequences in the immediate vicinity of the two alternative splice

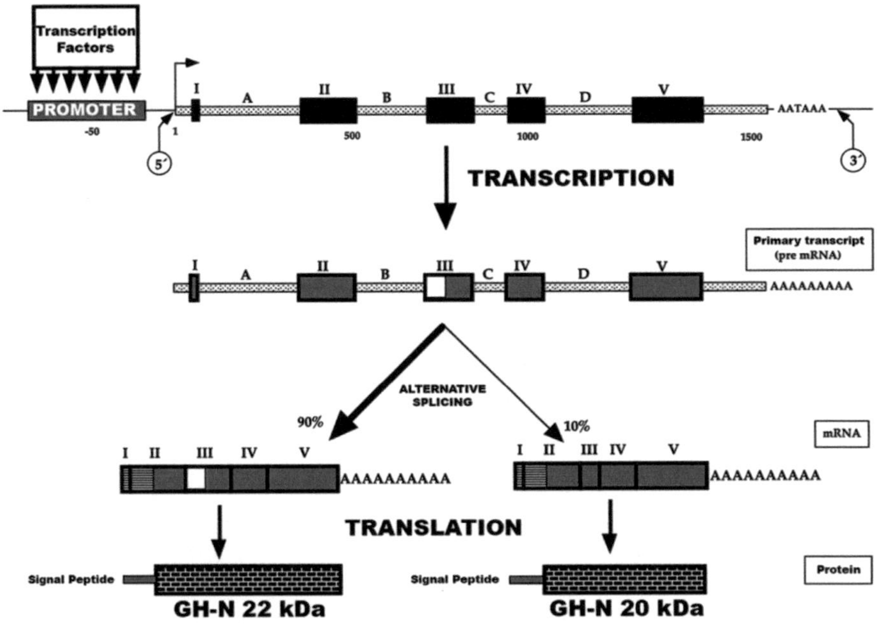

Figure 4.
Alternative splicing of the GH-N gene. After transcription of the GH-N gene, a part of intron III is cleaved (in 10% of transcripts, white rectangle), giving rise to a shorter GH-N form: The 20 kDA GH-N form.

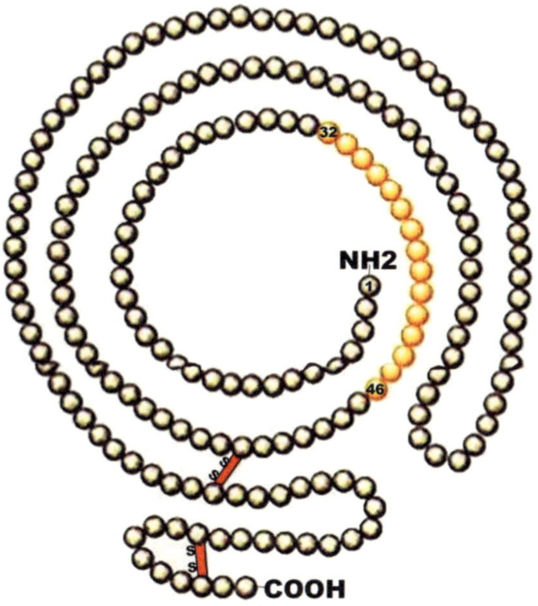

Figure 5.
Schematic description of the primary structure of GH-N 22 kDa. The molecule is a polypeptide formed by 191 amino acids arranged in a single chain in which there are two disulfide bridges (S–S) that join the cysteines located at positions 53 and 182 with those located at positions 165 and 189 (red lines). The alternative splicing removes amino acids 32–46 (represented by yellow circles) giving rise to the GH-N 20 kDa form.

acceptor sites, additional sequences further away in the transcript also contribute to this alternative splicing selection. These more distal sequences would not act individually, but instead would interact in such a way that the net level of alternative splicing in exon 3 would be dictated by the overall higher-order structure of the hGH-N transcript [32]. Curiously, a recent study concludes that although there is an important GH-N expression in retina, no splicing variants have been detected [33]. Our group found expression of GH and its receptor in hippocampal neural stem cells, but we did not differentiate whether the main GH form or the GH 20 kDa or both [34] were expressed. In the case of retina, it seems to be a particular situation, because both GH-N 22 kDa and 20 kDa bind and activate the same receptor (GHR); both forms exhibit similar, but not identical physiological activities, although the kinetics of signaling by GH-N 22 kDa and 20 kDa is different being weaker that of 20 kDa than of the major form, which can justify the different biological activities exhibited by these GH-N forms [35]. Furthermore, in humans, 20 kDa GH is a weaker agonist for the receptor of PRL than 22 kDa GH [36]. Other differences between both GH-N forms are related to their half-life in plasma. GH 20 kDa has a lower affinity for GH-binding proteins (GHBP) than GH 22 kDa [37] and fails to form a 1:1 complex with this binding protein [37]. The internalization rate of GH 20 kDa is less than that of GH 22 kDa [38]. Lastly, GH 20 kDa shows a stronger tendency to aggregate than GH 22 kDa [38–40]. The amplitude of GH secretory bursts is negatively correlated to percentage of body fat while testosterone plasma levels positively affect the secretory burst mass of the hormone [41]. Regarding the existence of the C-terminal disulfide bridge of GH, we know that it has been conserved throughout the evolution, although its role is unknown [42], but they seem to be fundamental for the maintenance of the active conformation of the hormone.

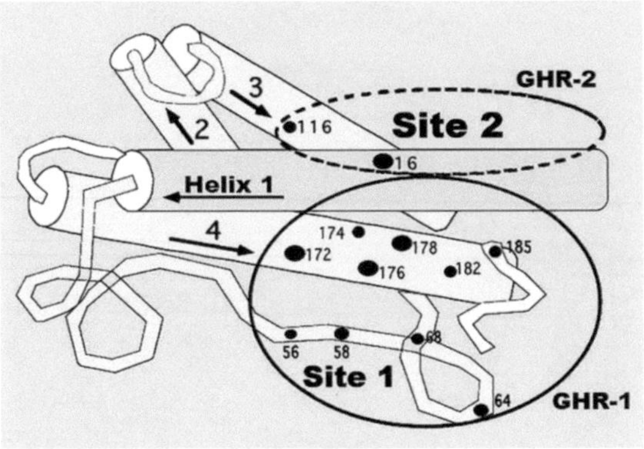

Figure 6.
Structure of 22 kDa GH-N showing the disposition of its 4 α-helix. Site 1 (outlined in a circle) is the one that first binds to the extracellular domain (ECD) of GHR-1, allowing the receptor to dimerize and Site 2 (outlined in a dotted line circle) of GH can bind to this second receptor (GHR-2). After that, the biological effects of the hormone begin.

GH has a complex tertiary structure, like other polypeptide hormones. Tertiary structure plays a role in how the hormone regulates receptor activation. GH is a long chain four α-helix bundle proteins (**Figure 6**).

A notable feature of their tertiary structure is that it contains no symmetry that might support equivalent binding environments for the receptors. How the two receptors bind to the asymmetric hormone was first revealed from the crystal structure of human growth hormone bound to the ECD of its receptor (hGHR). The characteristic arrangement of the four antiparallel α-helix is essential when it comes to produce the binding of GH to its receptor. Since this binding occurs in 1:2 ratio (one GH molecule and two receptor molecules) in each molecule of GH there are two receptor recognition epitopes, located at opposite ends of the nucleus of α-helix, site 1, and site 2. The structure shows that the two ECDs binding to site 2 and site 1, respectively, use essentially the same set of residues to bind to two sites on opposite faces of the hormone (**Figure 6**). This binding is characterized by extraordinary local and global plasticity at the binding surfaces. The two binding sites have distinctly different topographies and electrostatic character, leading to different affinities for the receptor ECDs. The high-affinity site, site 1, is always occupied first by ECD1. This sequence of events is required because productive binding of ECD2 at site 2 of the hormone requires additional contacts to a patch of the C-terminal domain of ECD1. The binding of ECD2 is the programmed regulatory step for triggering biological action, and it involves a set of highly tuned interactions among binding interfaces in two spatially distinct binding sites. The energetic relationships between the ECD1-ECD2 contacts and the hormone-ECD2 site 2 interactions are known to be important.

In the case of the 20 kDa GH-N, the reduced affinity exhibited by the receptor of the 22 kDa form suggests that conformational changes occurring in it affecting recognition epitopes.

As **Figure 2** shows, another form of GH, the GH variant or GH-V is expressed primarily in the placenta. This GH-V has an identical size that the pituitary 22 kDa GH-N, although both differ from each other by 13 amino acids dispersed throughout GH-V. However, the differences between GH-N and GH-V in terms of their somatogenic and metabolic activities are small. While secretion of pituitary GH-N is pulsatile under control from the hypothalamus, the secretion of placental GH-V is tonic and increases progressively in maternal blood during the second and third trimester [24].

Despite its placental expression, GH-V has been found in human testis [43–45], where most likely it plays an auto/paracrine function in reproduction (spermatogenesis). In addition to the testis it is likely, as indicated above, that GH-V, or some unknown gene related to it, can be expressed in other territories, including the pituitary.

2.3 The new GH-V variant

As with 22 kDa GH-N, the GH-V also undergoes alternative splicing during gene transcription, resulting from a 45 bp deletion produced by the use of an alternative acceptor site within exon 3 [46]. This group studied the effects of a 7-day treatment with 22 kDa hGH-N or 20 kDa hGH-V, administered subcutaneously (sc) in the same dose on the body composition and the endocrine and metabolic profiles in young male Wistar rats fed either with chow or a high-fat (HF) diet for 4 weeks post-weaning. Total body growth in the 20 kDa hGH-V-treated animals was intermediary between that of control and hGH-N-treated animals. Both 22 kDa hGH-N and 20 kDa hGH-V significantly reduced total body fat mass compared with control animals, and there were no differences between the GH isoforms in anti-lipogenic activity in animals fed the HF diet. Fasting plasma insulin and C peptide were significantly increased in animals on the HF diet and further increased by hGH-N but were unchanged in 20 kDa hGH-V-treated animals compared with saline-treated controls. Plasma volume was increased in hGH-N-treated animals but was unchanged in 20 kDa hGH-V-treated animals compared with controls. Furthermore, 20 kDa hGH-V had reduced lactogenic activity characteristic of hGH-N as tested in vitro compared with the 20 kDa hGH-N and 22 kDa hGH-N variants [46]. In summary, placental 20 kDa hGH-V retains some of the growth-promoting and all antilipogenic activities of pituitary 22 kDa hGH-N but has diminished diabetogenic and lactogenic properties compared with the native 22 kDa hGH-N [46]. These results indicate that some clear differences exist between 20 kDa GH-N and 20 kDa GH-V, perhaps indicating the different metabolic needs existing between not pregnant and pregnant women. A further very recent study was conducted to better characterize the in vivo activities of GHv (current name for this 20 kDa GH-V variant) in both sexes of a GH-deficient model of mice (GH−/− mice). GHv-treated GH−/− mice had significant increases to serum IGF-1, femur length, body length, body weight, and lean body mass and reduced body fat mass similar to mice receiving usual GH treatment. GH-N treatment increased circulating insulin levels and impaired insulin sensitivity; in contrast, both measures were unchanged in GHv-treated mice. The study also tested the ability of GH-N and GHv to stimulate the proliferation of human cancer cell lines and found that GHv has a decreased proliferative response in cancers with high PRLR (GHv is unable to bind to prolactin receptor) [5]. Their findings demonstrate that GHv can stimulate IGF-I and subsequent longitudinal body growth in GH-deficient mice similar to GH-N, but unlike it, GHv promoted growth without inhibiting insulin action and without promoting the growth of PRLR-positive cancers in vitro. Thus, GHv may represent improvements to current GH therapies especially for individuals at risk for metabolic syndrome or PRLR-positive cancers [5]. These results are surprising, especially if we think that this GHv comes from the placenta. What is the reason for showing such improved effects compared with those produced by the classic pituitary GH-N? Is GHv also expressed in tissues other than the placenta? Hopefully in the coming years we will have the answer to this question and, perhaps, many more surprises regarding this variant of GH, although it must be taken into account that these two studies have been carried out in two species, rats and mice, quite different from humans, including the GH that they produce.

3. The evolution of GH gene family

Growth hormone is a classic molecule in the study of the molecular clock, exhibiting a relatively constant rate of evolution in most orders of mammals, except primates and artiodactyls, where a dramatically enhanced rate of evolution (25–50-fold) has been reported. The rapid evolution of primate growth hormone occurred after the divergence of tarsiers (prosimians from Southeast Asia, with four extant species, living in tropical rain forests and included in the prosimian group, although some researchers see them as a link between prosimians and simians) and simians, but before the separation of old world monkeys (OWM) from new world monkeys (NWM). This event of rapid sequence evolution coincided with multiple duplications of the growth hormone gene, suggesting gene duplication as a possible cause of the accelerated sequence evolution. Multiple gene duplications and several gene conversion events both occurred in the evolutionary history of this gene family in OWM/hominoids. GHN genes in both hominoids and OWM are under strong purifying selection, while GHV genes in OWM and hominoids evolved at different evolutionary rates and underwent different selective constraints [47]. A key question is how hormone-receptor preferences have arisen among the duplicates, given that GH and PRL came from the same common gene from which they split 300–400 million years ago, moreover when both receptor genes (GHR and PRLR) show a surprising asynchrony in hormone and receptor gene duplications [48, 49].

Growth hormone exhibits a rare episodic pattern of molecular evolution characterized by sustained burst of rapid changes that are imposed very slow evolution on long periods. For example, there was a remarkable period of rapid change in the evolution of GH in primates or an ancestor and gave rise to the species-descending specificity [50, 51]. This pattern, also seen in placental growth hormones, are a consequence of selection, may reflect changes in the functions of GH additional to its basic growth-inducer effects [50, 51].

The biological specificity of GH is accompanied by significant differences in amino acid sequences, signifying an unusual episodic pattern of molecular evolution

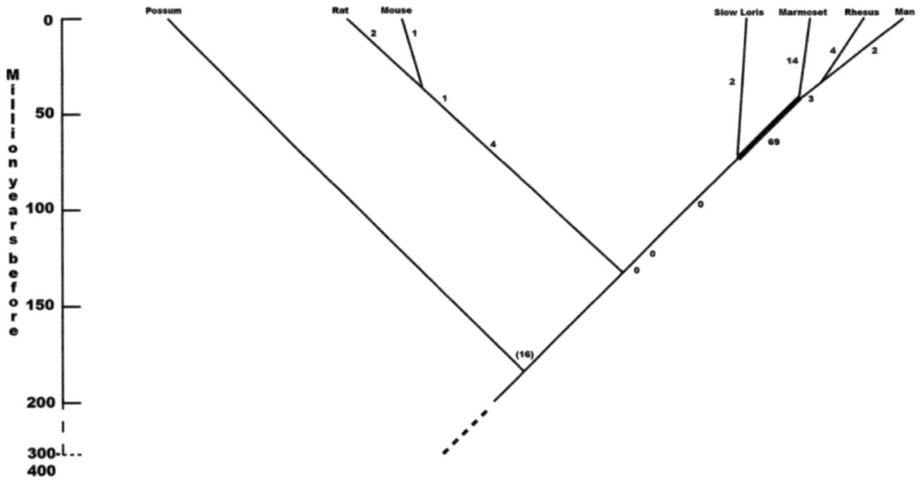

Figure 7.
Phylogenetic tree for mammalians GH (modified from ref. [53]). Y-axis represents million years in evolution; 0 is the current moment. Numbers of substitutions are indicated along the branches of the tree. Thick line indicates a period of rapid evolution for GH. There was another period of rapid evolution, which occurred earlier and gave rise to GH in alpaca, deer, and sheep, but this was omitted to simplify the figure, as a number of species were also omitted. Note the differences between rat and mouse, as well as between slow Loris, marmoset, and rhesus and man.

in which long periods wherein the sequence is markedly conserved (near-stasis) are interrupted by occasional bursts of rapid changes [50–53]. This is likely caused by positive Darwinian selection [52].

The burst of rapid changes on GH in the lineage leading to higher primates is particularly marked with substitution about 35% of all amino acid residues. This burst occurred before the separation of lineages leading to man and OWM (in fact, sequences of GHs of man and rhesus monkey are very similar) (**Figure 7**). However, there are not evidences for duplications of the GH gene in non-primate mammals.

The episode of rapid change seen for GH evolution in primates appears to be specific to the coding sequence for the mature protein hormone; the pattern of evolution seen for other components of the GH gene is different. Thus, when sequences of the signal peptide, 5′ untranslated region (**Figure 2**), or introns are analyzed, the burst of rapid evolution seen for the hormone is not present. The burst of change is specific to the protein-coding component of the gene indicating that its cause relates to the protein, resulting from adaptive change in response to selection, or loss of selective constraints due to loss of function. The episode of rapid acceleration is particularly marked for residues associated with receptor binding. There are evidences for an episode of rapid change in the GH receptor during primate evolution [53].

Of interest is the fact that regulatory sequences involved in the transcription of the gene are also affected by evolutionary changes. For example, a negative regulatory element (NRE3) is conserved in most mammals, including slow loris and marmoset. Two binding sites for the transcription factor Pit-1 are in the corresponding position for other mammalian GH genes [53]. It is notable that the distal and proximal sites in slow loris are much closer than in all other mammalian species, reflecting a deletion of about 14 nucleotides in the slow loris gene, which removes the site of the cyclic ANP response element existing in the GH gene. This region is not deleted in other mammalian GH genes, with the exception of marmoset and rabbit, but its different sequence makes it unlikely its functioning as a CRE [54]. A glucocorticoid response element, present in the first intron of the human GH gene, does not appear to be present in the marmoset or the slow loris gene. However, a number of putative negative thyroid hormone responsive elements exist in the human GH gene appear in the marmoset gene, but not in the slow loris gene. Therefore, these differences between these regulatory elements in human, marmoset, and slow loris GH genes indicate that the regulation of GH in these primate species shows significant differences. Of the many differences between non-primate and human GHs, that at position 169 (His in in pig and other non-primates, Asp at the equivalent position, 171, in man) appears to be most important in determining species specificity [55].

As we have seen, although in summary form, there are many changes in the molecular form of GH throughout the evolution, much more marked the more phylogenetically the species are apart from each other [56].

These changes seem to be related to the acquisition of new properties of GH, which leads the original gene to participate only in the regulation of growth. This would explain why GH has such high multiple actions in the body, rather than simply acting for growth, for which the initial gene seemed to emerge.

4. Conclusions

GH is a pleiotropic hormone produced mainly in the pituitary gland and the placenta, although the molecular forms of the hormone depend on the territory where they are produced. From an evolutionary point of view, GH comes from a common gene with PRL, from which it diverged millions of years ago, this may explain why GH and PRL share some physiological actions. All GH genes have five exons separated by four introns. Although GH has traditionally been considered a product of pituitary expression, today we know that the hormone can be produced in practically any territory and cell, the brain, retina, and gonads being particularly important. Currently, we can consider GH as a true prohormone, given the large number of forms that can be generated by proteolytic processing, both in the pituitary itself and in the tissues, processing that is tissue and sex-specific. Although it has been considered that only the classic GH-N can be generated in the pituitary, GH N-glycosylated products have been identified in this gland, suggesting that they may have similarities with placental GH-V. Regarding this, a new placental variant has been identified, GH-V2, which seems to play important roles, even antagonistic to those of pituitary GH-N. The original ancestral GH has undergone many structural changes throughout evolution, which has allowed the number of actions of the hormone to increase as the complexity of organisms increases. This leads to the fact that its initial effect as merely a growth hormone has been increased in a species-specific way.

Conflict of interest

The authors declare no conflict of interest.

Thanks

We want to thank the Foltra Foundation for the facilities given for the writing of this chapter. We also want to thank Diego de Souza (Infirmary, Foltra Medical Center) for his help in formatting the figures and references in this manuscript.

Author details

Jesús Devesa[1*] and Pablo Devesa[2]

1 Foltra Medical Center, TEO, Spain

2 Coordination and Research, School of Medicine, University Univalle, Cochabamba, Bolivia

*Address all correspondence to: jesus.devesa@usc.es

References

[1] Evans HM, Long JA. The effect of the anterior lobe administered intraperitoneally upongrowth maturation, and oestrus cycles of the rat. Anatomical Record. 1921;**21**:62-63

[2] Raben MS. Treatment of a pituitary dwarf with human growth hormone. The Journal of Clinical Endocrinology and Metabolism. 1958;**18**:901-903. DOI: 10.1210/jcem-18-8-901

[3] Hawco C, Houlden RL. Long-term follow-up of one of the first patients to receive human growth hormone therapy. AACE Clinical Case Reports. 2020;**6**:e262-e264. DOI: 10.4158/ACCR-2020-0278

[4] Dixon JS, Li CH. The amino acid composition of human pituitary growth hormone. The Journal of General Physiology. 1962;**45**(Pt 2):176-178. DOI: 10.1085/jgp.45.4.176

[5] List EO, Berryman DE, Basu R, Buchman M, Funk K, Kulkarni P, et al. The effects of 20 kDa human placental GH in male and female GH-deficient mice: An improved human GH? Endocrinology. 2020;**161**:bqaa097. DOI: 10.1210/endocr/bqaa097

[6] Hai-Nan L, Hui-Lin L, Zi-Qi Z, Gan L, Xue-Qi F, Xin Z. Cellular internalization and trafficking of 20 KDa human growth hormone. General and Comparative Endocrinology. 2019;**270**:82-89. DOI: 10.1

[7] Henry RK. Childhood growth hormone deficiency, a diagnosis in evolution: The intersection of growth hormone history and ethics. Growth Hormone & IGF Research. 2020;**55**:101358. DOI: 10.1016/j.ghir.2020.101358

[8] Schilbach K, Bidlingmaier M. Laboratory investigations in the diagnosis and follow-up of GH-related disorders. Archives of Endocrinology and Metabolism. 2019;**63**(6):618-629. DOI: 10.20945/2359-3997000000192

[9] Baş F, Abalı ZY, Toksoy G, Poyrazoğlu Ş, Bundak R, Güleç Ç, et al. Precocious or early puberty in patients with combined pituitary hormone deficiency due to POU1F1 gene mutation: Case report and review of possible mechanisms. Hormones (Athens, Greece). 2018;**17**(4):581-588. DOI: 10.1007/s42000-018-0079-4

[10] Hymer WC, Kennett MJ, Maji SK, Gosselink KL, McCall GE, Grindeland RE, et al. Bioactive growth hormone in humans: Controversies, complexities and concepts. Growth Hormone & IGF Research. 2020;**50**:9-22. DOI: 10.1016/j.ghir.2019.11.003

[11] Schlessinger J, Shechter Y, Willingham MC, Pastan I. Direct visualization of binding, aggregation, and internalization of insulin and epidermal growth factor on living fibroblastic cells. Proceedings of the National Academy of Sciences of the United States of America. 1978;**75**:2659-2663. DOI: 10.1073/pnas.75.6.2659

[12] Waters MJ, Rowlinson SW, Clarkson RW, Chen CM, Lobie PE, Norstedt G, et al. Signal transduction by the growth hormone receptor. Proceedings of the Society for Experimental Biology and Medicine. 1994;**206**:216-220. DOI: 10.3181/00379727-206-43745

[13] Lobie PE, Mertani H, Morel G, Morales-Bustos O, Norstedt G, Waters MJ. Receptor mediated nuclear

translocation of growth hormone. The Journal of Biological Chemistry. 1994;**269**:21330-21339

[14] Lobie PE, Wood TJ, Chen CM, Waters MJ, Norstedt G. Nuclear translocation and anchorage of the growth hormone receptor. The Journal of Biological Chemistry. 1994;**269**:31735-31746

[15] García-Barros M, Devesa J, Arce VM. Proteolytic processing of human growth hormone (GH) by rat tissues in vitro: Influence of sex and age. Journal of Endocrinological Investigation. 2000;**23**:748-754. DOI: 10.1007/BF03345065

[16] Singh RN, Seavey BK, Lewis UJ. Heterogeneity of human growth hormone. Endocrine Research Communications. 1974;**1**:449-464. DOI: 10.3109/07435807409089000

[17] Seeburg PH. The human growth hormone gene family. Nucleotide sequences show recent divergence and predict a new polypeptide hormone. DNA. 1982;**1**:239-249. DOI: 10.1089/dna.1.1982.1.239

[18] Cooke NE, Ray J, Watson MA, Estes PA, Kuo BA, Liebhaber SA. Human growth hormone gene and the highly homologous growth hormone variant gene display different splicing patterns. The Journal of Clinical Investigation. 1988;**82**:270-275. DOI: 10.1172/JCI113582

[19] Lewis UJ, Dunn JT, Bonewald LF, Seavey BK, Vanderlaan WP. A naturally occurring structural variant of human growth hormone. The Journal of Biological Chemistry. 1978;**253**:2679-2687

[20] Hirt H, Kimelman J, Birnbaum MJ, Chen EY, Seeburg PH, Eberhardt NL. The human growth hormone gene locus: Structure, evolution, and allelic variations. DNA. 1987;**6**:59-70. DOI: 10.1089/dna.1987.6.59

[21] Diaz MJ, Dominguez F, Haro LS, Ling N, Devesa J. A 12-kilodalton N-glycosylated growth hormone-related peptide is present in human pituitary extracts. The Journal of Clinical Endocrinology and Metabolism. 1993;**77**:134-138. DOI: 10.1210/jcem.77.1.8325936

[22] García-Barros M, Costoya JA, Ríos R, Arce V, Devesa J. N-glycosylated variants of growth hormone in human pituitary extracts. Hormone Research. 2000;**53**:40-45. DOI: 10.1159/000023512

[23] Baumann G. Growth hormone heterogeneity: Genes, isohormones, variants and binding proteins. Endocrine Reviews. 1991;**121**:424-449. DOI: 10.1210/edrv-12-4-424

[24] Baumann GP. Growth hormone isoforms. Growth Hormone & IGF Research. 2009;**19**:333-340. DOI: 10.1016/j.ghir.2009.04.011

[25] Spolaore B, Polverino de Laureto P, Zambonin M, Fontana A. Limited proteolysis of human growth hormone at low pH: Isolation, characterization, and complementation of the two biologically relevant fragments 1-44 and 45-191. Biochemistry. 2004;**43**:6576-6586. DOI: 10.1021/bi049491g

[26] Such-Sanmartín G, Bosch J, Segura J, Gutiérrez-Gallego R. Generation of 5 and 17 kDa human growth hormone fragments through limited proteolysis. Growth Factors. 2009;**27**:255-264. DOI: 10.1080/08977190903110121

[27] Luna M, Martínez-Moreno CG, Ahumada-Solórzano MS, Harvey S, Carranza M, Arámburo C. Extrapituitary growth hormone in the chicken

reproductive system. General and Comparative Endocrinology. 2014;**203**:60-68. DOI: 10.1016/j.ygcen.2014.02.021

[28] Lewis UJ, Ling N, Haro LS. Variants forms of human growth hormone in serum. Acta Paediatrica. 1994;**399**(Suppl):29-31

[29] Sinha YN, Jacobsen BP. Human growth hormone (hGH)-(44-191), a reportedly diabetogenic fragment of hGH, circulates in human blood: Measurement by radioimmunoassay. The Journal of Clinical Endocrinology and Metabolism. 1994;**78**:1411-1418. DOI: 10.1210/jcem.78.6.8200944

[30] Lewis UJ, Lewis MJ, MAM S, Staten NR, Galosy SS, Krivi GG. A recombinant-DNA-derived modification of human growth hormone (hGH 44-191) with enhanced diabetogenic activity. Molecular and Cellular Endocrinology. 1991;**78**:45-54. DOI: 10.1016/0303-7207(91)90184-t

[31] Rowlinson SW, Waters MJ, Lewis UJ, Barnard R. Human growth hormone fragments 1-43 and 44-191: in vitro somatogenic activity and receptor binding characteristics in human and non-primate systems. Endocrinology. 1996;**137**:90-95. DOI: 10.1210/endo.137.1.8536647

[32] Estes PA, Urbanek M, Ray J, Liebhaber SA, Cooke NE. Alternative splice site selection in the human growth hormone gene transcript and synthesis of the 20 kDa isoform: Role of higher order transcript structure. Acta Paediatrica. Supplement. 1994;**399**:42-47. DOI: 10.1111/j.1651-2227.1994.tb13285.x

[33] Pérez-Ibave DC, Garza-Rodríguez ML, Pérez-Maya AA, Rodríguez-Sánchez IP, Luna-Muñoz M, Martínez-Moreno CG, et al. Expression of growth hormone and growth hormone receptor genes in human eye tissues. Experimental Eye Research. 2019;**181**:61-71. DOI: 10.1016/j.exer.2019.01.011

[34] Devesa P, Agasse F, Xapelli S, Almengló C, Devesa J, Malva JO, et al. Growth hormone pathways signaling for cell proliferation and survival in hippocampal neural precursors from postnatal mice. BMC Neuroscience. 2014;**15**:100. DOI: 10.1186/1471-2202-15-100

[35] Yao-xia L, Jing-yan C, Xia-lian T, Ping C, Min Z. The 20 kDa and 22 kDa forms of human growth hormone (hGH) exhibit different intracellular signalling profiles and properties. General and Comparative Endocrinology. 2017;**48**:49-54. DOI: 10.1016/j.ygcen.2017.04.010

[36] Tsunekawa B, Wada M, Ikeda M, Uchida H, Naito N, Honjo M. The 20-kilodalton (kDa) human growth hormone (hGH) differs from the 22 kDa hGH in the effect on the human prolactin receptor. Endocrinology. 1999;**140**:3909-3918. DOI: 10.1210/endo.140.9.6959

[37] Baumann G, Stolar MW, Amburn K, Barsano CP, BC DV. A specific growth hormone-binding protein in human plasma: Initial characterization. The Journal of Clinical Endocrinology and Metabolism. 1986;**62**:134-141. DOI: 10.1210/jcem-62-1-134

[38] Wada M, Uchida H, Ikeda M, Tsunekawa B, Naito N, Banba S, et al. The 20-Kilodalton (kDa) human growth hormone (hGH) differs from the 22 kDa hGH in the complex formation with cell surface hGH receptor and hGH-binding protein circulating in human plasma. Molecular Endocrinology. 1998;**12**:146-156

[39] Wada M, Ikeda M, Takahashi Y, Asada N, Chang KT, Takahashi M, et al. The full agonistic effect of recombinant

20 kDa human growth hormone (hGH) on CHO cells stably transfected with hGH receptor cDNA. Molecular and Cellular Endocrinology. 1997;**133**:99-107

[40] Chapman GE, Rogers KM, Brittain T, Bradshaw RA, Bates QJ, Turner C, et al. The 20,000 molecular weight variant of human growth hormone: Preparation and some physical and chemical properties. The Journal of Biological Chemistry. 1981;**256**:2395-2401

[41] Veldhuis JD, Liem AY, South S, Weltman A, Weltman J, Clemmons DA, et al. Differential impact of age, sex steroid hormones, and obesity on basal versus pulsatile growth hormone secretion in men as assessed in an ultrasensitive chemiluminescence assay. The Journal of Clinical Endocrinology and Metabolism. 1995;**80**:3209-3222

[42] Junnila RK, Wu Z, Strasburger CJ. The role of human growth hormone's C-terminal disulfide bridge. Growth Hormone & IGF Research. 2013;**23**:62-67

[43] Untergasser G, Kranewitter W, Walser F, Madersbacher S, Dirnhofer S, Berger P. The testis as eutopic production site of human growth hormone, placental lactogen and prolactin, possible autocrine/paracrine effects on testicular function. Wiener Klinische Wochenschrift. 1996;**108**:541-546

[44] Untergasser G, Kranewriter W, Schwarzler P, Madersbacher S, Dirnhofer S, Berger P. Organ-specific expression pattern of the human growth hormone/placental lactogen gene-cluster in the testis. Molecular and Cellular Endocrinology. 1997;**130**:43-60

[45] Berger P, Untergasser G, Hermann M, Hittmair A, Madersbacher S, Dirnhofer S. The testis-specific expression pattern of the growth hormone/placental lactogen (GH/PL) gene cluster changes with malignancy. Human Pathology. 1999;**30**:1201-1206

[46] Vickers MH, Gilmour S, Gertler BH, Breier K, Tunny K, Waters MJ, et al. 20 kDa placental hGH-V has diminished diabetogenic and lactogenic activities compared with 22 kDa hGH-N while retaining anti lipogenic activity. American Journal of Physiology. Endocrinology and Metabolism. 2009;**297**:E629-E637

[47] Ye C, Li Y, Shi P, Zhang Y-p. Molecular evolution of growth hormone gene family in old world monkeys and hominoids. Gene. 2005;**350**:183-192

[48] Ocampo Daza D, Larhammar D. Evolution of the growth hormone, prolactin 2 and somatolactin family. General and Comparative Endocrinology. 2018;**264**:94-112

[49] Nicoll CS, Mayer GL, Rusell SM. Structural features of prolactin and growth hormones that can be related to their biological properties. Endocrine Reviews. 1986;**7**:169-203

[50] Wallis M. The molecular evolution of vertebrate growth hormones: A pattern of near-stasis interrupted by sustained bursts of rapid change. Journal of Molecular Evolution. 1996;**43**:93-100

[51] Wallis M. Function switching as a basis for bursts of rapid change during the evolution of pituitary growth hormone. Journal of Molecular Evolution. 1997;**44**:338-350

[52] Ohta T. Pattern of nucleotide substitutions in growth hormone-prolactin gene family: A paradigm for evolution by gene duplication. Genetics. 1993;**134**:1271-1276

[53] Krawczak M, Chuzhanova NA, Cooper DN. Evolution of the proximal

promoter region of the mammalian growth hormone. Gene. 1999;**273**:143-151

[54] Wallis OC, Wallis M. Cloning and characterization of the rabbit growth hormone-encoding gene. Gene. 1995;**163**:253-256

[55] Behncken SN, Rowlinson SW, Rowland JE, Conway-Campell BL, Monks TA, Waters MJ. Aspartate 171 is the major primate-specific determinant of human growth hormone. Engineering porcine growth hormone to activate the human receptor. The Journal of Biological Chemistry. 1997;**272**:27077-27083

[56] Wallis OC, Zhang Y-P, Wallis M. Molecular evolution of GH in primates: Characterization of the GH genes from slow Loris and marmoset defines an episode of rapid evolutionary change. Journal of Molecular Endocrinology. 2001;**26**:249-258

Chapter 3

Pituitary Growth Hormone Secretion and Cell Growth Hormone Production: Regulation of Their Secretion and Their Signaling Pathways

Jesús Devesa and Pablo Devesa

Abstract

Growth hormone (GH) performs very diverse functions in the organism, and this is the reason by which the regulation of the secretion of this hormone is very complex; although the primary regulators are growth hormone-releasing hormone (GHRH) and somatostatin, it is in turn regulated mainly by adrenergic and cholinergic pathways, and other factors can act directly on its secretion, particularly on the somatostatin, thus affecting the pituitary secretion of GH. In this chapter, we will analyze the transcription of GH gene and how GH release is affected by different neurotransmitters, metabolic substrates, feeding and fasting, and other hormones, placing special emphasis on why pituitary secretion of GH is sexually dimorphic.

Keywords: GH, GHRH, somatostatin, Pit-1, neurotransmitters and GH secretion, metabolic substrates and GH secretion, hormonal regulation of GH

1. Introduction

Growth hormone (GH) is produced mainly in the anterior pituitary gland (AP), but its expression also takes place in many other territories, such as the brain, the gonads, retina, adrenals, where this hormone, or its derivatives formed by enzymatic cleavage, performs very diverse and specific roles, although how GH is regulated at these levels is not well known.

In the anterior pituitary gland (AP), GH is the most abundant hormone, and it is synthesized in eosinophilic cells, the somatotrophs developed due to the action of the transcription factor Prop-1 (homeobox protein prophet of Pit-1), whose expression leads to the pituitary development of Pit-1-specific lineages. The Pit-1 t transcription factor determines the specific cellular expression of GH (and also prolactin (PRL) and thyroid-stimulating hormone (TSH)).

GH can be found in the pituitary as early as 8 weeks of fetal development, being a key factor for the development and maturation of multiple organs and tissues,

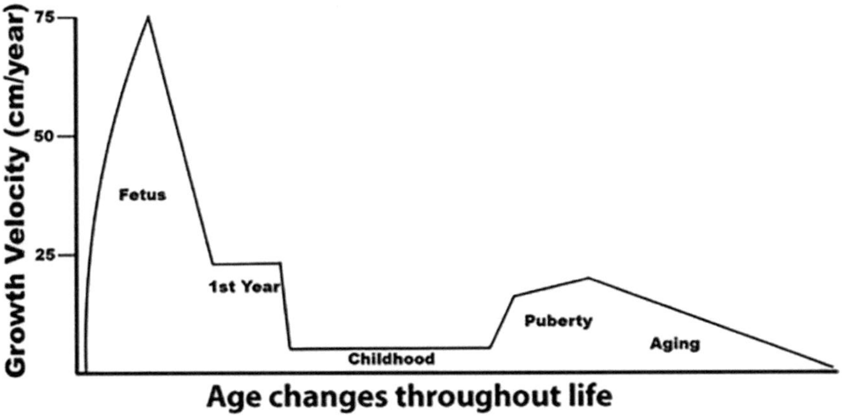

Figure 1.
Changes in growth velocity throughout life. Note that when puberty approaches and sex steroids increase, they act on the synthesis and release of pituitary GH, so the growth velocity increases with respect to childhood. However, from the end of puberty, GH secretion progressively decreases with aging.

such as the brain. This fetal GH does not seem to play any role in the growth of the fetus, despite the high growth velocity during the fetal period. In fact, anencephalic fetuses are of normal height at birth. A high growth rate also occurs after birth, up to 6–10 months of age, after which growth slows down; these changes after birth are shown in **Figure 1**.

Growth hormone secretion is pulsatile in nature; this is related to the optimal induction of physiological effects at the peripheral level: target tissues for GH appear to be more sensitive to the frequency at which this hormone arrives during a period of time than to the total amount of GH secreted during a similar period [1]. A clear pulsatility of GH only appears after birth; the majority of the GH pulses are associated with slow wave sleep, a period in which the amplitude of GH secretion reaches the maximum [2]. However, sleep processes do not appear to exert a predominant influence on GH release, since when sleep and circadian processes are misaligned, the blunting of the sleep-related GH pulse is counteracted, as in individuals deprived of sleep, by a compensatory mechanism that promotes GH pulses during wakefulness [3].

GH secretion is sexually dimorphic from the puberty, due to gender-related free estradiol (fE_2) levels in the brain [4]. In humans, plasma fE_2, but not testosterone (T), levels were shown to be strongly correlated with total and pulsatile GH release rates [5]. However, it cannot be ruled out whether there is an imprinting effect of sex steroids on the hypothalamic structures that govern the underlying hypothalamic-somatotrophic rhythm (HSR), as occurs in rat [6]. In women, the episodic secretion of GH is more frequent, but of a lower amplitude, while plasma GH values during trough periods are slightly higher than in men. Therefore, the fraction of GH secreted in peak episodes is lower in women, although it changes according to the phase of the menstrual cycle once puberty is established, reaching higher values in the late follicular phase of the menstrual cycle.

2. Regulation of pituitary GH secretion

In 1985, Plotsky and Vale demonstrated in rats that the episodic secretion of GH was under a dual hypothalamic control exerted by growth hormone-releasing hormone

(GHRH) and somatostatin (SS) [7]. According to this model, the determinant of pulsatile GH secretion would be the rhythmic alternate release of hypothalamic GHRH and SS. The existence of a similar intrinsic hypothalamic-somatotroph rhythm (HSR) that governs the pituitary GH secretion was also demonstrated by our group in normal humans [8]. Intrinsic HSR most likely remains functionally active throughout life. However, plasma GH levels are highly age-dependent and show a different pattern of secretion in men and women. Both age and gender physiologically modify the standard pattern of activity, as plasma GH changes reflect. In addition, a number of factors, such as neurotransmitters, other hormones, and metabolic intermediates (amino acids, glucose, and free fatty acids), mainly acting at on the hypothalamic or pituitary level, or both, but also on the pituitary, can modify, both acutely and chronically, the pattern of GH release. Therefore, GH control has traditionally been classically considered as extraordinarily complex, although its complexity has increased over time due to the discovery of new factors that contribute significantly to modify the pituitary secretion of GH, as we will see more ahead. This also reflects the complexity of the multiple actions that GH exerts in the organism, far beyond its role in growth.

2.1 GHRH

Reichlin [9] was the first to propose the existence of a hypothalamic GHreleasing factor. However, it took several years to isolate and characterize this putative factor [10, 11]. The initial isolation of GHRH was facilitated by the existence of human tumors ectopically producing growth hormone-releasing activity [12, 13].

GHRH is a peptide that belongs to the brain-intestinal peptides family, the glucagon secretin family. Two main GHRH forms of GHRH, composed of 40 and 44 amino acids, have been described in the human hypothalamus, occurring as a result of the alternative processing of RNA during the transcription of a GHRH gene located on chromosome 20. These are the result of alternative processing of RNA from the transcript of a single GHRH gene located on chromosome 20. Once it was demonstrated that the GH-releasing activity of GHRH was located between amino acids 1–29, it was possible to produce shorter forms of the peptide (e.g., GHRH-29, known as GRF-29) to be used in clinical applications, such as a provocative test. Since the removal of amino acids 30–44 does not affect the GH-releasing activity of the peptide, shorter forms (e.g., GRF-29) have been synthesized for clinical purposes. On the contrary, modifications at the N-terminal end of the molecule markedly decrease its biologic activity. The half-life of GRF-29 in plasma is very short, about 2 minutes, as it is very rapidly inactivated by circulating proteases. This has been led to the development of potent GHRH agonists and antagonists., as it will be shown later.

GHRH-producing neurons involved in GH control are located primarily in the arcuate nucleus of the hypothalamus from where they reach the median eminence and release the peptide into the primary plexus of the hypothalamic-pituitary portal vascular system. In the pituitary, GHRH binds to its receptor on somatotrophs and induces a rapid, dose-dependent release of GH, but also the transcription of the GH gene as well as the differentiation, growth, and proliferation of these cells.

In addition to this, GHRH-producing cells have also been found in other hypothalamic and extra-hypothalamic structures. These cells, however, do not project to the median eminence but to several other hypothalamic and extrahypothalamic regions. This suggests that apart from its GH-releasing activity, the peptide also plays a neuromodulator role in the central nervous system. In addition, there is GHRH production in numerous tissues and organs, such as the gonads.

GHRH is encoded by a gene located in chromosome 20 (20q11.2), which consists of five exons separated by four introns. Exon I is not coding; exon II codes for the signal peptide and a small N-terminal connecting peptide; exon III encodes the majority of the mature GHRH molecule (including the biologically active peptide); exon IV encodes most of the C-terminal peptide; and exon V encodes the rest of the C-terminal peptide and also contains 3′ untranslated sequences [14] (**Figure 2**).

A variant of the GHRH precursor has been described. It is formed by RNA processing that excludes three nucleotides at the beginning of exon V, leading to the absence of serine-103 in the precursor molecule. This precursor will undergo identical processing to that of the 108 amino acids precursor, with the only difference that in this case the C-terminal peptide will be made up of only 30 amino acids [14].

An interesting aspect, little clarified, is how the transcription of the GHRH gene is regulated. KO mice for the homeobox transcriptional factor Gsh-1 exhibit a dwarf phenotype and abolished GHRH expression [15], and Gsh-1 binds on multiple binding sites of GHRH gene promoter. Co-expression of the transcription factor conditional cAMP response element binding (CREB) protein significantly enhances Gsh-1-induced expression of the GHRH gene, suggesting a cooperative role for the coactivator protein. The Gsh-1 homeobox protein is key for the expression of the GHRH gene.

And what is the role of CREB in the transcription of GHRH gene? In mice in which CREB is lost in the brain, but not in the pituitary, the amount of GHRH peptide is reduced, indicating that CREB is required for efficient production of GHRH in the hypothalamus [16]. This would explain the relationships between CREB and Gsh-1 for hypothalamic GHRH transcription.

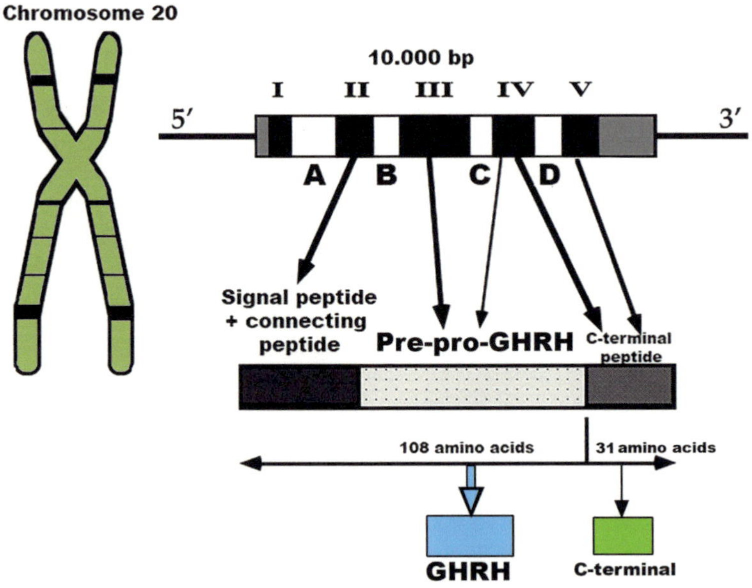

Figure 2.
Structure of the GHRH gene and transcription of GHRH. As the figure shows, the gene-encoding GHRH is composed by five exons (I–V) and four introns (A–D). GHRH is synthetized as a precursor (pre-pro-GHRH) formed by 108 amino acids (including the signal peptide), which is proteolytically cleaved and gives rise to the mature GHRH (blue rectangle) together with a C-terminal peptide formed by 31 amino acids. As the figure shows, the precursor of GHRH is produced by exon III and a small portion of exon IV (black arrows 3 and 4).

Another GHRH gene transcription factor is the nuclear factor of activated T cells (NFAT). Site-directed mutagenesis experiments demonstrated the direct binding of NFAT at five sites of the GHRH promoter, suggesting that NFAT is involved in the depolarization-induced (with high potassium) transcriptional activation of GHRH gene in the hypothalamic neurons producing it [17].

Apart of its hypothalamic production, GHRH is a peptide produced in many human tissues and organs, where it plays a very important physiological and etio-pathological role in addition to its main role in the synthesis and release of GH.

2.2 Somatostatin (SS)

At the end of the 1960s, McCann's group observed that bovine hypothalamic extracts were able to inhibit the release of GH from pituitary cultures. Five years later, Guillemin' group [18–20] reported the identification of the factor responsible for that effect: GH-release-inhibiting factor or SS. Several studies soon demonstrated that SS was not only a GHrelease-inhibiting factor but also a strong inhibitor of a number of secretory events, endo- or exocrine, whose analysis is beyond the objective of this chapter.

Although SS is a tetradecapeptide (SS-14), a form SS-28 has been found in pituitary portal blood, also capable of inhibiting pituitary GH release. Structurally, SS is a tetradecapeptide, although a 28-amino-acid form is also present in pituitary portal blood. Both molecular forms apparently have similar GH-release-inhibiting activity.

SS-producing neurons involved in GH control are found mainly are mainly located in the anterior periventricular and paraventricular nuclei of the hypothalamus. Their axons project to the median eminence, where SS is released into the hypothalamic-pituitary portal circulation to antagonize in a dose-dependent manner the GH-releasing effect of GHRH, and also, although in less degree, the proliferative activities of this GH-releasing peptide.

As with GHRH, SS is synthesized in the form of a precursor, pre-pro-somatostatin, a 116 amino acid peptide (including the 24 amino acid signal peptide) whose biologically active portion is at the C-terminal end. The gene that codes for SS is located on chromosome 3 and consists of two exons separated by an intron [21, 22]. In it, a cAMP response sequence is found between nucleotides −48 and −41 [23]. The transcription of the gene is stimulated by CREB, phosphorylated after an increase in cAMP levels, and the subsequent activation of protein kinase A (PKA). The coding sequence has a high degree of conservation in exons.

The proteolytic processing of pro-somatostatin gives rise to two physiologically important variants, SS-14 and SS-28. SS-14 is a tetradecapeptide that is formed after cleavage by endoproteases of the precursor between amino acids 101 and 102, while SS-28 corresponds to the sequence of the SS-14 form with an N-terminal extension of 14 amino acids [24] (**Figure 3**). The proteolytic cleavage of pro-somatostatin in the basic domain (Arg/Lys) or the monobasic domain (Arg) of the C-terminal end gives rise to SS-14 and SS-28, respectively (**Figure 3**). Both molecular variants are encoded by exon 2.

Both SS-14 and SS-28 are widely distributed in the body, being particularly abundant throughout the brain, but also in the gastrointestinal tract (particularly in the pancreas and stomach), which may be related to its inhibitory functions of endocrine and exocrine secretions. SS also acts as a neurotransmitter and regulator of the immune system. Most likely, these are the reasons for the existence of five receptors (SST1 to SST5) with different localizations and biological activities. Since the fact that SS has anti-proliferative and anti-angiogenic effects and given that its half-life is very

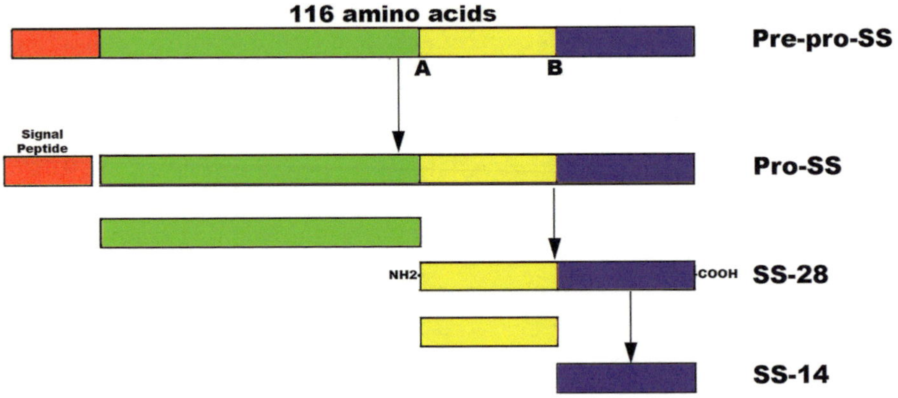

Figure 3.
Proteolytic processing of pre-pro-somatostatin. Proteolytic processing of pre-pro-somatostatin releases the signal peptide (formed by 24 amino acids) and then the connection peptide (green rectangle, formed by 64 amino acids). As indicated SS-28 is formed by the 14 amino acids extension (yellow rectangle) to the N-terminal end of SS-14. A: monobasic domain. B: basic domain.

short (1–3 min), as well as the wide spectrum of its biological responses, led to the need to develop selective stable receptor agonists that are bind strongly to selective SS receptor subtypes, therefore, in addition to blocking GH secretion (e.g., treatment of Acromegaly), these SS agonists are being used for the therapy of neuroendocrine tumors, diabetic complications (nephropathy, retinopathy), anti-neoplastic therapies. More recently, it has been shown that these agonists have anti-inflammatory and anti-nociceptive effects [25].

Before reviewing how GHRH and SS interact in the control of pituitary synthesis and release, we will describe how the pituitary receptors for these hormones are stimulated. Most likely this is different than how they are stimulated in other tissues in which they are expressed.

2.2.1 Pituitary GHRH receptor

The GHRH receptor (GHRHR) is a 423 amino acid protein (including the signal peptide) that belongs to a subfamily of G-protein-coupled receptors [26, 27]. Its producing gene is located to the short arm of chromosome 7 (7p13–p21) near the epidermal growth factor receptor (EGFR) gene [28].

The gene has a very complex structure [26]. Unlike other G-protein-coupled receptors, which do not have introns, in this gene there is a coding sequence interrupted by several introns. An alternative ribonucleic acid (RNA) processing has been described that results in the insertion of 41 amino acids just before the beginning of the sixth transmembrane domain. This type of processing does not appear to be exclusive to GHRHR [29], although its physiological significance is unknown.

Like the rest of the receptors that use G proteins for intracellular signaling, the GHRHR has seven transmembrane hydrophobic domains, linked to each other by six loops, three intra-cytoplasmatic and three extracellular.

The N-terminal extracellular domain of GHRHR contains a site for N-glycosylation as well as six cysteine residues and an aspartate residue that are conserved in this receptor family, while the third intracellular loop and C-intracellular domain contain several potential phosphorylation sites, which may regulate signaling and receptor internalization.

GHRHR mRNA has been detected in the pituitary, gonads, placenta, kidney, hypothalamus, and many brain regions, although it is expressed predominantly in the anterior pituitary due to its dependence of Pit-1 (also known as GHF-1).

GHRH interaction with its GHRHR on the somatotroph leads to the activation of the associated G_s protein, allowing the α_s subunit to separate from the dimer ßγ. The activation of G_s protein will then have a double effect: (1) It induces the release of GH stored in secretory granules by allowing the entry of Ca^{++} into the cell, and (2) the α_s subunit activates the adenylyl cyclase (AC), which leads to increased cellular levels of cAMP [26, 30, 31]. This, in turn, activates the protein kinase A (PKA), which induces a cascade of phosphorylation and activation of the transcription factor CREB [32]. The activation of CREB induces the pituitary-specific transcription factor Pit-1 [33–35]. Therefore, Pit-1 represents a pathway by which GHRH can increase GH synthesis in the pituitary. Interestingly, Pit-1 also regulates GHRHR synthesis, as if there were some kind of feedback positive between them, and also induces its own expression. In fact, the GHRHR is not expressed in the pituitary of dw/dw mice that lack functional Pit-1 [30]. In turn, the non-activation of CREB leads to somatotroph hypoplasia and dwarfism in mice.

Other signaling pathways activated after the GHRH-GHRHR interaction involve the Phospholipase C/Inositol phosphatide/Protein kinase C (PLC/IP/PKC) pathway, which is perhaps also responsible for the increase of Ca^{++} for GH secretion in somatotrophs. In this case, the Ca^{++} would come from the endoplasmic reticulum. This effect would occur through the stimulation of phospholipase C (PLC) through the complex $G_{\beta\gamma}$ of heterotrimeric proteins G. Activated PLC produces diacylglycerol (DAG) and inositol triphosphate (IP3) that leads to the release of Ca^{++} from the endoplasmic reticulum. However, this effect has not been seen in some inferior species [36];

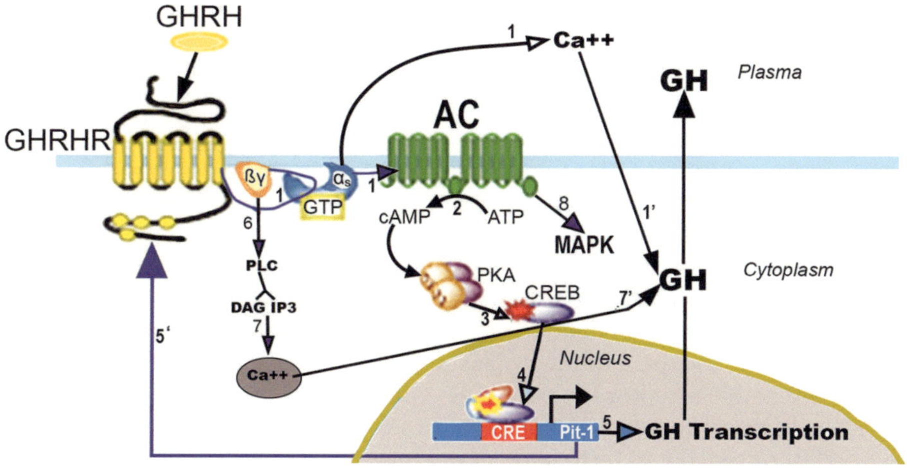

Figure 4.
Interaction GHRH-GHRHR at the pituitary level and synthesis and secretion of GH. After interacting with its receptor, GHRH activates the associated G_s protein, and the α_s subunit is separated from the ßγ dimer (1). This allows: (1) the entry of Ca++ into the cell (1′), and the activation of adenylyl-cyclase (AC) (1) leading to increased formation of cAMP from ATP (2). cAMP activates protein kinase A [PKA] that phosphorylates CREB (3). Phosphorylated CREB translocates to nucleus (4) and induces Pit-1, which, in turn, leads to the transcription of the GH gene (5) and also regulates the synthesis of GHRHR (5′). On the other hand, the $G_{\beta\gamma}$ complex of heterotrimeric proteins G activates PLC (6) producing DAG and IP3 (7) that leads to the release of Ca^{++} from the endoplasmic reticulum (7′) for the secretion of GH. Another mediator of the effects of GHRH is the mitogen-activated protein kinase (MAPK) pathway, which may mediate the effects of GHRH on the proliferation of somatotrophic cells (8).

another signaling pathway observed in a wide variety of cell types as a mediator of GHRH effects is the mitogen-activated protein kinase (MAPK) pathway, which may mediate the effects of GHRH on somatotroph cell proliferation [37], although it cannot be ruled out that the proliferative effect is mediated by *c-fos* activated by CREB. These pathways activated by the interaction GHRH-GHRHR are shown in **Figure 4**.

2.2.2 Pituitary SS receptor

The wide variety of actions of SS and the different organs and tissues in which this hormone acts explain the existence of different types of receptors for this exocrine and endocrine inhibitory hormone. The five types of receptors for SS (SSTR1 to SSTR5) belong to the DRY family of G-protein-coupled receptors, whose common characteristic is the presence of an Asp-Arg-Tyr sequence, located near the boundary between the third transmembrane domain and the second intracellular loop, which is very important for transmission of the SS signal [38, 39]. Each of the receptors is encoded by a different gene, located on different chromosomes, and their coding region lacks introns (as is the case with most G-protein-coupled receptors).

SS receptors show a homology between them that varies between 45% and 61%, and 100 residues are the same in all of them. The highest degree of homology occurs in the sequences corresponding to the seven transmembrane helices, while the NH2 and COOH ends are the most divergent both in sequence and in length [39].

The SSTR2 receptor is the prototype of the SS receptors and is the one that mediates the inhibition that SS exerts on GH secretion in the pituitary gland [40]. This receptor has the same affinity for SS-14 as for SS-28 [39].

The binding of SS with its receptor induces the activation of the related G protein (**Figure 5**), which in the case of the SSTR2 receptor is a Gi protein. The activation of the Gi protein will produce an inhibition of the activity of AC, a reduction in the entry of Ca^{++} through voltage-gated channels, and the appearance of rectifying potassium currents [39, 41], which leads to hyperpolarization of the somatotropic membrane. The net product of these actions would be the inhibition of the transcription of CREB-dependent genes [42], which counteracts the stimulatory effect of GHRH. However, in the control of GH, SS acts basically by inhibiting the release of the hormone, while its effect on the synthesis of GH would be of minor importance. SS also exerts anti-proliferative effects on mediated somatotrophs, through the activation of a tyrosine phosphatase dependent on the SSTR2 receptor [43, 44].

SS receptors undergo desensitization phenomena after prolonged exposure to agonists [39], which seems to depend on the phosphorylation of serine and threonine residues, located in the third intracytoplasmic loop, by the action of the enzyme BARK (ß-adrenergic receptor kinase) [39].

Once the basic components that generate a secretory rhythm of GH have been established, the task should be to understand how this rhythm may begin, how it can be modulated by factors external to said system, and what could be the relationships between its basic components: GH, GHRH, and SS. That is, is there a signal associated with GHRH that causes the interruption of SS release or is SS release linked to the termination of GHRH secretion, as initially thought? Is the end of GHRH secretion directly dependent on the increase in GH produced by the peptide, or is there a GH-associated signal that stimulates SS release and blocks GHRH secretion?

Before answering these questions, there is the need to analyze how the pituitary transcription of GH is regulated by transcription factors, as well as the structural organization of Pit-1, the main one of these factors. After that, the mechanisms that

DOI: http://dx.doi.org/10.5772/intechopen.108382

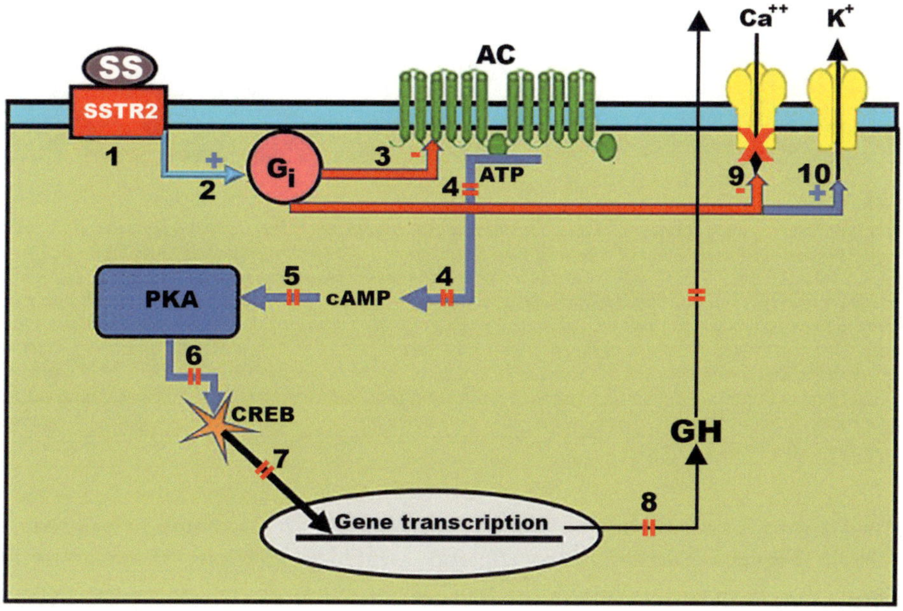

Figure 5.
Interaction of SS-SSTR2 at the pituitary level and its effects on GH secretion. In the pituitary gland, binding of SS to its pituitary receptor SSTR2 (1) activates (2) the Gi-related inhibitory protein. This leads to inhibition (3, –) of AC. Consequently, ATP cannot give rise to cAMP (4, =). The lack of the necessary cAMP prevents (5, =) the activation of PKA, which then cannot initiate the phosphorylation cascade (6, =) that would activate CREB, and non-phosphorylated CREB cannot translocate to the cell nucleus (7, =) to initiate the transcription of specific genes, including GH (8, =). Then, the secretion of GH would be blocked (=). Furthermore, activated Gi blocks (9, –) the channels involved in the entry of Ca^{++} into the cell (X). This Ca^{++} is necessary for the release of GH. Moreover, activated Gi leads to the appearance (10, +) of rectifying potassium currents, which leads to hyperpolarization of the somatotropic membrane. Overall, SS blocks GH secretion and partially affects GH synthesis. +: stimulates; –: inhibits; =: interrupts or cuts; **X***: disrupts; blue arrows: stimulates. Red arrows: inhibits.*

regulate the hypothalamic secretion of GHRH and SS will be reviewed. Given that they are neurosecretory products, it seems logical that neurotransmitters play the main role in such regulation.

2.3 Regulation of the transcription of the gene GH-N

Pit-1 is the most important of the transcription factor involved in controlling the expression of the GH-N gene. It is a highly conserved protein, belonging to the family of transcription factors with POU domains that are a subclass of the homeobox genes involved in cell development and differentiation processes [42, 45]. In man, the gene encoding Pit-1 is located on chromosome 20 and is made up of six exons and five introns.

Pit-1 is not only expressed in the pituitary but also in many other territories where it may be involved in the control of cell proliferation. In the pituitary, Pit-1 not only controls the expression of GH and PRL, but also that of the ß chain of TSH, that of the GHRH gene, and, curiously, that of its own gene. Pit-1 is also, as we have already described, a key factor for the development, differentiation, and survival of somatotrophs. The absence of Pit-1, due to mutations, produces alterations in cell development and hormonal synthesis [46, 47]. On the contrary, the overexpression of this factor has been related to the appearance of pituitary adenomas [48, 49].

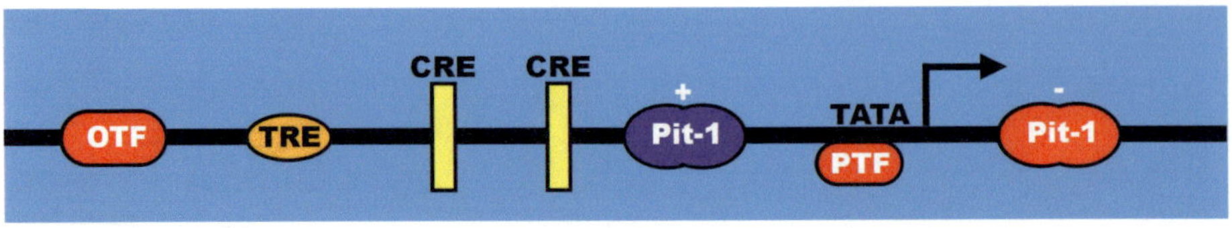

Figure 6.
Schematic structure of the promoter of Pit-1. The human Pit-1 promoter is Pit-1-dependent and autoregulated. Activating sequence of basal Pit-1 self-transcription (blue figure, +) is located at position −55 bp in the promoter, while a second sequence of Pit-1, in this case inhibitor of self-transcription (red figure, −), is located at position +15 bp. The transcriptional activity is negatively regulated by Oct-1 and mediated by the octamer-binding site OTF (red figure). Intracellular levels of cAMP positively affect Pit-1 transcription acting through two cAMP-responsive elements (CRE9 localized in the proximal promoter region). PTF: pituitary transcription factor, a cell-type-specific TATA element that seems not to be conserved in the human promoter. TRE: TPA-responsive element, located at position −490 in the proximal promoter region, whose possible function could be to mediate a complete transcriptional shut-off of the human Pit-1 gene when the DNA-binding activity of the Pit-1 protein is inhibited by mitotic phosphorylation. (TPA: 12-O-tetradecanoylphorbol-13-acetate responsive element).

Basal transcription of the Pit-1 gene is self-regulated by two binding sequences for Pit-1 itself (**Figure 6**): an activating sequence located at position −55 and an inhibitory sequence, located at position +15. In addition, the transcription of this gene would be regulated by factors capable of increasing the levels of cAMP and activating cAMP response-element-binding (CREB). Likewise, the activation of the protein kinase C-dependent signaling pathway seems to play an important role, having described the existence of a response element to AP-1 (*activating protein-1*), which exerts an inhibitory effect on transcription [50].

Three isoforms of Pit-1 have been described, of which two of them with molecular weights of 31 kDa and 33 kDa are jointly called Pit-1 or Pit-1a and represent the majority of Pit in the pituitary. The third isoform, Pit-2 or Pit-1b, appears by alternative RNA processing; it retains its ability to bind DNA and may be involved in the differentiation process of somatotrophs [51].

Structurally, three regions are distinguished in Pit-1: (1) a DNA-binding zone located at the C-terminal end, in which two domains can be differentiated, a homeodomain (POU_{HD}) of 60 amino acids found throughout the family of transcription factors with POU domains, and a specific POU domain (POU_s) of 75 amino acids located at the N-terminus of the POU_{HD} domain. The POU_{HD} domain is the one that comes into contact with the DNA and is the one that contains the Pit-1 nuclear transfer signal. For its part, the POU_s domain does not directly contact the DNA, but surely increases the binding of POU_{HD} to it, stabilizing the DNA-protein complex [42]. (2) A region responsible for the activation of transcription, which comprises the 72 amino acids from the N-terminal end of the molecule and binds to site 1 of the response element. (3) A “bridge” region located between the two above.

In addition to Pit-1, GH transcription in man is increased by GHRH and glucocorticoids and inhibited by SS and activin [46]. GHRH increases intracellular cAMP that produces an increase in the synthesis of CREB proteins that bind to the two CREs (proximal and distal) of the GH promoter (**Figure 7**) increasing the transcription of the gene. On the contrary, both SS and activin counteract these actions by inhibiting the synthesis of cAMP. In the case of glucocorticoids, their effect is produced by the direct action of their receptor on the response elements located at the promoter and first exon level (**Figure 7**), facilitating the access of other transcription factors (especially Pit-1 and CREB) to their specific binding sequences [42].

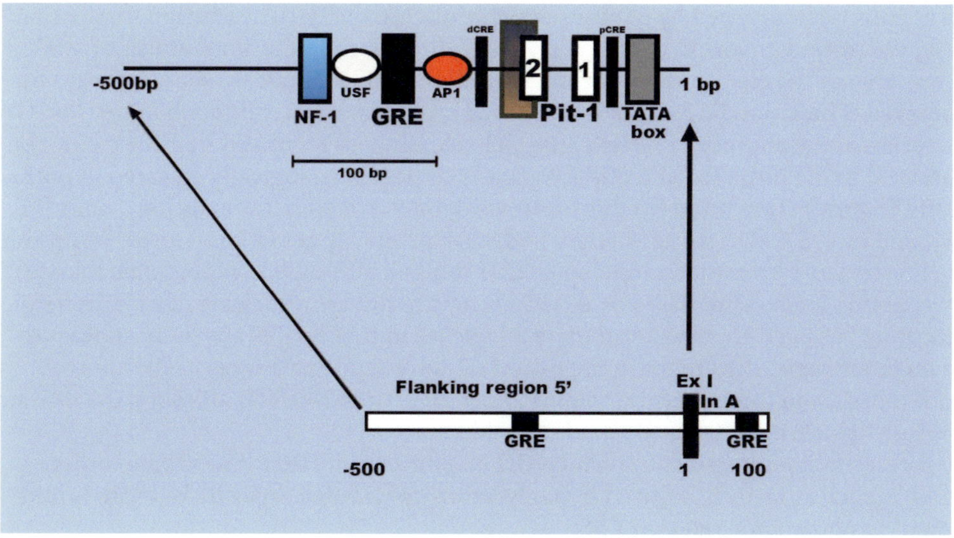

Figure 7.
Structural organization of the GH promoter. The figure shows the location of the DNA response elements for the different transcription factors of the GH-N gene. GRE: glucocorticoid response element. CRE: cAMP response elements (dCRE: distal and pCRE: proximal). 1 and 2 represent, respectively, the binding sites for the DNA-binding domain and for the transactivation domain of the specific transcription factor Pit-1. NF-1 (nuclear factor-1) represents the binding sequence of the silencers of said family (found in rat). Sp-1 (specificity protein 1) provides GHRH responsiveness in the GH gene through activation of adenylate cyclase. The lower part of the figure shows the location of the second GRE at intron A of the gene. Ex: Exon. In: Intron. bp: base pairs.

2.4 Neurotransmitter regulation of GHRH and SS secretion

Our group demonstrated that GH responses to an exogenous GHRH challenge and other stimuli for GH secretion, in humans, were strongly dependent on the functional status of the intrinsic HSR at the time of testing [8]. From this it is concluded that: (a) GHRH was not able to interrupt HSR the hypothalamic rhythm that governs GH secretion, and (b) stimuli capable of altering this rhythm must act mainly by inhibiting the release of SS, rather than by stimulating endogenous GHRH. In fact, a continuous infusion of GHRH does not alter the pulsatile GH secretion. The maintenance of pulsatile GH secretion when GHRH is continuously infused supports these conclusions.

On this basis, the role of the main neurotransmitters in the control of GH secretion was reevaluated.

2.4.1 Adrenergic pathways and GHRH and SS secretion

Central adrenergic pathways play an important role in the control of GH in mammals [52]. Pharmacological blockade of catecholamine synthesis or release leads to the abolition of pulsatile GH secretion, an effect reversed by the administration of α_2-adrenergic receptor agonists. Conversely, pharmacological blockade of α_1-adrenergic receptors does not modify the latter phenomenon [53].

For years, experimental evidence led to the postulate that GH secretion induced by central α_2-adrenergic stimulation was mediated by increased GHRH release.

However, consistent with our observations described above [8], GH responses to any challenge must depend on the following: (a) potentiation of endogenous GHRH release, while SS tone is physiologically low, or (b) a primary inhibitory effect on

SS release, accompanied by a secondary stimulation of GHRH secretion. In the first case, the responses would be sporadic, depending on the functional status of HSR at the time of the test. In the second case, a significant GH release should always be observed. The rationale for this is also based on the fact that, although there is no GH secretion in the absence of GHRH, the GH-releasing effect of this peptide cannot be observed in the presence of a physiologically or pharmacologically elevated hypothalamic SS production tone. On this basis, we demonstrated in humans [54], dogs [55, 56], and rats [57] that α2-adrenergic pathways primarily act in GH control by inhibiting SS release, rather than inducing GHRH release, although hypothalamic release of this peptide is also stimulated by α2-adrenergic pathways. Blockade of α2-adrenergic receptors has been shown to stimulate SS release in rabbits [58], so there appears to be no interspecies difference in the role of α2-adrenergic pathways in GH control. Both the α1- and β-adrenergic systems antagonize the α2-effect, although the former appears to have no physiological relevance in man.

β-Adrenergic antagonists enhance GH responses to GHRH [59, 60] and other stimuli, such as insulin-induced hypoglycemia [59], which logically is due to inhibition of hypothalamic release of SS.

In summary, the adrenergic system performs two antagonistic functions in the regulation of human GH: facilitator, mediated by α2-adrenergic receptors that act mainly by inhibiting the release of SS and secondarily by inducing the release of GHRH; and inhibitory, dependent on β-adrenergic activity, which stimulates SS secretion and inhibits GHRH release [61, 62].

2.4.2 Dopaminergic pathways and GHRH and SS secretion

Studies on the role of dopaminergic pathways in GH secretion have yielded contradictory results; stimulant and inhibitory effects have been described. Vance et al. [63] reported that dopamine (DA) increased GH secretion caused by GHRH in humans, suggesting an inhibitory effect of DA on the hypothalamic release of SS, exerted at the level of the median eminence. However, administration of centrally acting DA agonists, such as bromocriptine [63] and CV 205–502 [64], has also been shown to enhance both spontaneous GH secretion and GHRH-induced release. Therefore, DA would presumably inhibit SS release at least at two levels (the hypothalamus proper and the median eminence).

In vitro studies have indicated that DA stimulates SS secretion from the median eminence; furthermore, an increase in SS release in the pituitary portal blood has been observed after intraventricular injection of DA in rats [54]. On the other hand, our group has reported that central DA receptor blockade with metoclopramide was able to enhance the GH response to GHRH during periods of physiologically increased delivery of SS to the pituitary, but not when the SS tone appeared to be low [65]. This suggested a stimulatory role for DA in hypothalamic SS release, but also that this could be mediated by an indirect mechanism: negative DA modulation of NA release. This hypothesis is not inconsistent with the aforementioned studies [63, 64] showing that DA agonists centrally facilitate GH release. In fact, centrally acting DA agonists, such as bromocriptine, induce a biphasic pattern of GH secretion: inhibition followed by rebound stimulation.

In summary, everything suggests that the central role of DA in the control of GH in humans depends mainly on its effects on adrenergic transmission to SS neurons. Therefore, DA would act as a modulator rather than a direct regulator of GH secretion.

2.4.3 Cholinergic pathways and GHRH and SS secretion

Evidence indicates that cholinergic pathways play a key role in the control of GH secretion in both humans and laboratory animals [54, 66, 67]. Inhibition of central cholinergic pathways with muscarinic receptor-blocking drugs (atropine, pirenzepine) strikingly decreases GH release induced by a number of physiologic and pharmacologic stimuli [68, 69]; however, these have no effect on the GH secretion elicited by insulin-induced hypoglycemia. Conversely, enhancement of cholinergic tone with muscarinic cholinergic agonists potentiates GHRH-stimulated GH secretion and stimulates basal GH release as well.

It has been hypothesized that cholinergic synapses may be the final pathway for a variety of stimuli inducing GH secretion. The administration of anti-SS antibodies blocks the inhibitory effect of atropine on GHRH-elicited GH release in normal rats. Moreover, the acetylcholinesterase inhibitor pyridostigmine increases GH responses to a maximal stimulating dose of GHRH even when there is an abnormally high SS tone, as in obesity. Therefore, the facilitating role of cholinergic pathways on GH secretion appears to be mediated by an inhibition of hypothalamic SS release. In fact, while muscarinic cholinergic agonists stimulate basal GH secretion and its response to GHRH, muscarinic antagonists inhibit both responses [68, 69], immunoneutralization of GHRH by a GHRH-specific antibody does not affect the secretion of GH induced by pyridostigmine [70].

Because α_2-adrenergic and muscarinic cholinergic pathways appear to play an equally important role in GH neuroregulation in humans, both by inhibiting hypothalamic SS release, it has been investigated the functional relationships between these two neurotransmitter pathways. Our findings obtained showed that while α_2-adrenergic activation was able to overcome the inhibitory effect of muscarinic cholinergic blockade on GHRH-induced GH secretion, blockade of α_2-adrenergic receptors counteracted the stimulating action of the pyridostigmine on said response [61]. Given the above, we speculated that in the control of GH, the α_2-adrenergic neurons are located distally to the cholinergic neurons, so that the latter contribute to GH secretion by modulating the functional activity of the former, as occurring in peripheral tissues. The fact that known stimuli for SS inhibition acting through adrenergic neurons (e.g., galanin and insulin hypoglycemia) are able to overcome the inhibitory effect of atropine on GH secretion is consistent with this theoretical scheme, which, moreover, does not exclude the possibility that the cholinergic input directly reaches SS neurons and inhibits them, as acetylcholine does in hypothalamic cultures *in vitro*.

Thus, the degree of cholinergic activity likely determines the amount of catecholamines released into the synaptic cleft of SS neurons. This, in turn, activates either SS-stimulating β-adrenergic receptors (responsive to low concentrations of catecholamines) or SS-inhibiting α_2-adrenergic receptors (responsive only to high catecholamine concentrations). To better understand the actions of these three neurotransmission pathways on GH secretion, see **Figure 8**.

2.4.4 Other neurotransmitters acting on GHRH and SS secretion

Given the complexity of hypothalamic structures and the relationships between different neurotransmitters in the control of hypothalamic functions, it seems logical that neurotransmitters other than those analyzed may play a role in the control of the hypothalamic discharge of GHRH and SS release into portal blood. This is the case for serotonin, gamma-aminobutyric acid (GABA), nitric oxide (NO), and endogenous opioids.

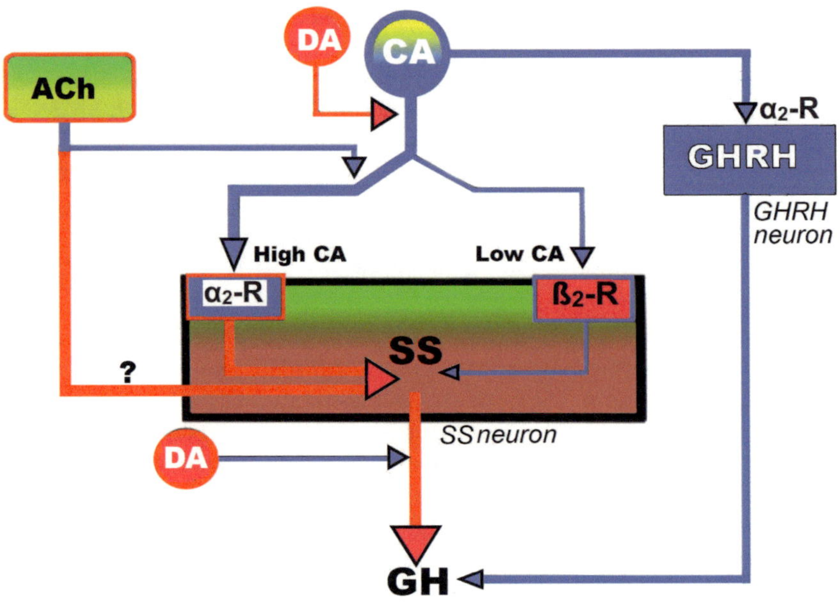

Figure 8.
Main control of GH by neurotransmitters. The main role is played by adrenergic signals (CA) to somatostatin-producing neurons (SS). Depending on the amount of CA released into the synaptic cleft, inhibitory α_2-R receptors (high CA input) or stimulatory β_2-R adrenoceptors (low CA supply) are activated in SS neurons (see text). This, in turn, would be positively modulated by ACh supply to CA neurons, but negatively by DA. However, a direct inhibitory effect of ACh on SS neurons, or a stimulatory effect of DA on SS release in the median eminence (?), cannot be excluded. On the other hand, the stimulatory effect of CA on GHRH release, mediated by stimulation of α_2-adrenergic receptors in the GHRH neuron, appears to be less important than the inhibition of SS. Therefore, SS appears to be the main determinant of the secretion pattern of GH. Blue arrows: stimulation. Red arrows: inhibition.

2.4.4.1 Serotonin

In rats, serotonin stimulates GH secretion, whereas in humans, stimulation or inhibition or even no effect has been described [53]. These discrepancies could be due to the low specificity of the drugs used, but also to the fact that serotonin acts on multiple types of receptors [71]. However, serotonin fibers have been found in periventricular nucleus and the arcuate nucleus where the SS and GHRH-producing neurons are located, respectively, so this neurotransmitter must play some role in the control of GH secretion. Indeed, administration of an anti-GHRH antibody suppresses serotonin-induced GH secretion in rats [72] and hypoglycemia-induced GH release in man is abolished by administration of serotonin antagonists [73]. In turn, the administration of L-tryptophan, a precursor of serotonin, increases the release of GH.

2.4.4.2 Gamma aminobutyric acid (GABA)

GABA is an important neurotransmitter whose actions at the central level are mainly inhibitory. In 1984, GABA was reported to modulate the secretion of AP hormones secretion by acting at both the hypothalamic and pituitary levels [74]. Intraventricular administration of GABA in rats stimulated LH and GH release. This effect was attributed to the fact that GABA induced an elevation of hypothalamic NA and median eminence DA levels as well as AP DA levels. The authors concluded that GABA plays a physiological role in the control of AP hormone secretion, primarily through hypothalamic action [74]. These data supported a previous study in which a dual action of GABA was postulated in the control of the hormonal secretion of the AP gland: one mediated *via* the central nervous system and the other exerted directly at the level of the AP [75]. Inhibition of GABA degradation and blockade of GABA transmission as well as administration of GABA and GABA mimetic drugs have all been shown to affect GH secretion. However, there are many controversial findings. For example, ancient studies indicated that GABA inhibited the release of SS [76, 77]. The effects of GABA may depend on the site of action within the hypothalamic-pituitary axis and the hormonal milieu. What is clear is that GABA participates in the regulation and actions of a very important GH secretagogue, such as ghrelin. In addition, long-term GHRH administration increases GABA levels in all brain regions [78], and GHRH activates GABA receptors in the cerebral cortex [79]. In all, these data suggest that GABA participates in the control of GH secretion but acts as a modulator of the actions of other neurotransmitters or peptides.

2.4.4.3 Nitric oxide (NO)

The gaseous neurotransmitter nitric oxide (NO) is synthesized from arginine by the action of nitric oxide synthase (NOS). NO performs a number of very diverse actions in the body, both at the level of the central nervous system and in the periphery. NO plays a stimulatory role *in vivo* and *in vitro* on GH secretion in acromegalic patients [80]. In rats, NO stimulates secretion of GHRH therefore increasing secretion of GH; however, it was also found that GHRH, in turn, increases production of NO in somatotrophs blunting GH secretion [81]. In dogs, inhibition of NO blunts GHRH-induced GH secretion, suggesting that NO acts by decreasing hypothalamic SS release [82]. However, years before, Aguila [83] argued that the hypothalamic increase in NO induced by GHRH would act on SS neurons increasing both the synthesis and release of SS. More recent studies conducted in fetal human primary cell cultures indicated that NO stimulated GH secretion, probably *via* the cGMP pathway [84]. This is consistent with further data demonstrating that a strong GH secretagogue, such as ghrelin, requires the activation of the NOS/NO pathway, and its subsequent GC/cGMP signal transduction pathway, as essential steps to induce GH secretion from somatotrophs [85].

In summary, it appears that NO plays a role in mediating the GH response to neuroendocrine factors, but its role appears to change depending on environmental conditions, including mitochondrial functioning, either in the hypothalamus or in the pituitary gland itself.

2.4.4.4 Endogenous opioids

There are three major types of endogenous opioid peptides that can be considered as neurotransmitters: endorphins, enkephalins, and dynorphins, which are respectively derived from three different precursor proteins: pro-opiomelanocortin (POMC), preproenkephalin A and B [86], although this number increased after the discovery of endomorphin-1 and endomorphin-2 [87].

The endogenous opioid peptides play many different physiological and pharmacological effects in humans, including neuroendocrine actions. In the case of the control of GH secretion, it has been shown that β-endorphin, administered intravenously or directly into the rat hypothalamus, increases GH secretion, an effect mediated by α_2-adrenergic pathways, *via* stimulating GHRH and inhibiting SS release, because β-endorphin antiserum reduced the stimulatory effect of clonidine on GH release [88]. Treating rats with an antiserum against GHRH inhibits the GH stimulatory response to β-endorphin [89], and the inhibitory effect of SS on GH secretion is antagonized by endogenous opioids [90].

At this point, there is the need to remark that exogenous non-peptide opioids have different effects on GH secretion when administered to man. In general, whereas acute opioid administration increases GH secretion, the effects of chronic opioid administration are much more complex. For example, intrathecal administration of opioids in chronic pain patients inhibits GH secretion, but the response is affected by sex, body composition, and insulin resistance [91].

2.5 Regulation of GH by metabolic substrates

GH is a metabolic hormone that plays an important role in regulating carbohydrate, fat, and protein metabolism in humans. Therefore, it is logical that metabolic substrates play a modulator role in the control of GH secretion.

2.5.1 Glucose and GH secretion

Glucose is a nutrient that the brain uses almost exclusively to provide energy [92], so glycemic regulation is key to maintaining normal brain function.

Basically, acute hyperglycemia inhibits GH secretion, either in basal conditions or in response to a number of stimuli acting on the central nervous system. Conversely, the acute decrease in plasma glucose concentrations leads to a rapid GH discharge. GH is one of the four counter-regulatory hormones that oppose the hypoglycemic actions of insulin (the other three are catecholamines, glucagon, and cortisol).

The opposite effects of hypo- and hyperglycemia seem to take place in the hypothalamus in SS-producing neurons. Hypoglycemia leads to the inhibition of SS secretion, while hyperglycemia induces SS release as cholinergic agonism prevents hyperglycemic blockade of GHRH-induced GH secretion [53]. The increase in GH induced by hypoglycemia appears to be almost totally refractory to cholinergic blockade [53]. Since hypoglycemia strongly activates adrenergic transmission, an inhibition of SS release mediated by α_2-adrenoceptors [54, 61] would favor GH secretion, also positively acting on the secretion of GHRH. The former is most likely dependent on increased glucose-induced SS release, as cholinergic agonism prevents hyperglycemic blockade of GHRH-induced GH secretion [53]. The latter, in turn, appears to be mainly due to by inhibition of SS secretion. The fact that the increase in GH induced by hypoglycemia appears to be almost totally refractory to cholinergic blockade [53] does not invalidate such a possibility. Since hypoglycemia strongly activates adrenergic transmission, an inhibition of SS release mediated by α_2-adrenoceptors [54, 61] would favor GH secretion. Furthermore, as seen above, an enhancement of endogenous GHRH release would also take place. In fact, in mice, it has been seen that hypoglycemia activates GHRH neurons [93]. However, also in mice, it has been reported that hypoglycemia produces a decrease in GH release [94, 95], most likely mediated by specific activation of neurons producing Neuropeptide Y [96]. Therefore, it is not yet clear, at least in mice, what is the role of glucose detection in GHRH neurons in controlling GH release during hypoglycemia due to the possible involvement of other central systems, such as the neurons that produce NPY.

As it is logical, hyperglycemia has to act in the opposite way to hypoglycemia. Consequently, as described, hyperglycemia increases the hypothalamic release of SS [97] and GHRH-induced GH secretion drastically decreases [98]. Furthermore, GH content in the pituitary is decreased in diabetic rats.

However, GH responses to diabetes are different in rats and humans. In humans, type I diabetes presents with increased pulsatile GH secretion [99, 100], perhaps because there is a greater pituitary sensitivity to GHRH [101]; however, in type II diabetes, GH secretion is negatively affected [102–104]. These differences are most likely due to the different fat mass in both types of diabetes and its consequences on insulin secretion, suggesting that insulin may be a regulator of GH release in both types of diabetes.

2.5.2 Free fatty acids (FFA) and GH secretion

FFA can totally block GH responses to a number of stimuli, including that of exogenous GHRH [105]; therefore, it has been postulated that the inhibition of GH release induced by FFA is mediated by an increased secretion of hypothalamic SS, although the data obtained *in vitro* point to a direct pituitary inhibition also [106]. In humans, the GH response to combined administration of pyridostigmine plus GHRH is markedly altered when plasma FFAs are increased by administration of a lipid-heparin infusion [107]. This favors the possibility of a main inhibitory action of FFAs on somatotrophs, which could be exerted by modifying the bilayer structure of the membranes and, therefore, affect their ability to detect stimulating signals (e.g., GHRH) or activate subsequent secretory mechanisms (GHRH or increased intracellular Ca^{2+}). Interestingly, the response of other pituitary hormones to their hypothalamic releasing factors appears to be unaffected by the increase in plasma FFA in humans. Therefore, the inhibitory action of FFAs on GH secretion appears to be very specific. This is curious, because an important metabolic effect of GH is to induce the release of FFA from adipose tissues [108, 109]. Obese subjects show increased plasma FFA levels, while GH secretion is practically abolished, and if we suppress circulating FFA with antilipolytic drugs, obese patients regain normal GH secretion [110, 111]. Elevated levels of FFA are associated with insulin resistance and insulin hypersecretion; therefore, the decrease in GH in obesity may be caused by altered insulin feedback, rather than being a consequence of elevated plasma FFA levels, although there is no doubt about the effect of FFAs on the pituitary response to GHRH and SS. The opposite situation is seen in fasting: GH secretion increases and stimulates FFA release [108, 109].

In total, these relationships between GH and FFA appear to be related to the need to maintain metabolic homeostasis.

2.5.3 Amino acids and GH secretion

Given the clear and important anabolic effects that GH exerts on protein synthesis and metabolism, it is at least surprising that only two basic amino acids, arginine and ornithine, exert a powerful effect on GH secretion in humans. Arginine clearly increases the maximal GH responsiveness to GHRH in humans [112, 113], even when an elevated SS tone seems to exist (e.g., in the elderly). Furthermore, pyridostigmine does not potentiate GH response to arginine [114]. Hence, the inhibition of hypothalamic SS release appears to be the mechanism involved in the GH-releasing effect of this amino acid. Furthermore, arginine proceeds from citrulline and can be transformed into NO by the action of NOS, which suggests that another mechanism of action of this amino acid in GH secretion may be secondary to the increase in NO synthesis at the central level.

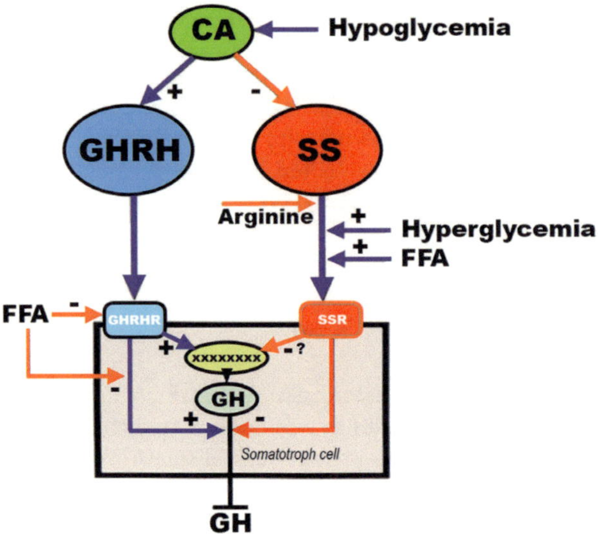

Figure 9.
Basic control of GH by metabolic substrates. Perhaps the main regulator is plasma glucose levels, given the effects of GH as a counterregulatory hormone. Hypoglycemia is detected by glucose-sensitive neurons and leads to the release of catecholamines (CA). These have a dual effect: they inhibit the release of SS (red arrow, −) and stimulate the release of GHRH (blue arrow, +). In somatotrophic cells, there are receptors for GHRH (GHRHR) and SS (SSR) that when stimulated induce respectively the synthesis and release of GH (GHRH, blue arrow, +) or the inhibition of the secretion of this hormone (SSR, red arrow, −), although in the case of SSR, it cannot be ruled out that they also act negatively on the synthesis of pituitary GH (?). The effects of FFA seem to be more varied, facilitating the release of SS, like hyperglycemia (blue arrow, +), and also acting at the level of somatotrophs, inhibiting the activation of GHRH and/or preventing the stimulating effect of GHRH on GH release (red arrow, −). Arginine appears to directly inhibit SS release (red arrow, −), allowing GHRH release. Although not depicted in the figure, the pituitary SS receptor is type 2, SSTR2.

The effect of arginine on GH secretion is much more powerful than that of ornithine.

The basic effects of metabolic substrates on GH secretion are schematically depicted in **Figure 9**.

2.6 Regulation of GH by other hormones and peripheral and central peptides

Endocrine products mainly related to GH control are those that play a general permissive role in the body, such as thyroid hormones and glucocorticoids, or those related to maturational changes, such as sex steroids, but also other hormones, such as Insulin and IGF-I play an important role in the control of GH synthesis and release. Furthermore, as with other pituitary hormones, GH secretion is self-regulated. In addition, as GH plays multiple important roles in practically all the tissues of the body, especially of a homeostatic metabolic regulatory nature, a series of central and peripheral peptides, some of which can be considered as authentic hormones, clearly participate in the modulation of GH synthesis and secretion, as we will analyze in this section.

2.6.1 Thyroid hormones and GH secretion

Thyroid hormones are permissive hormones needed for a physiological response to any other hormone. Hypothyroidism is known to be associated with impaired linear growth, which is easily normalized by the administration of replacement doses of thyroid hormones. Consistent with it, there is a decrease in plasma levels of IGF-I, which are normalized after the treatment with thyroid hormones. Both spontaneous GH secretion and GH responses to classic stimuli are markedly reduced in clinical hypothyroidism in humans or in induced hypothyroidism in rats [115]. There are several explanations for these facts. First, the lack of thyroid hormones strongly affects the hypothalamic synthesis of GHRH [116]. Although this directly leads to impaired GH synthesis, additional effects of hypothyroidism may depend on the lack of both the positive modulation that thyroid hormone induces on the number of GHRH-receptor sites in somatotrophs and its facilitatory role on GHRH binding to its receptors. Furthermore, GH-secretory mechanisms are also impaired in hypothyroidism, not only because a decrease in the hypothalamic content of GHRH [116, 117] but also because SS appears to be increased, at least in the rat [117–119].

Here at this point, it is of interest to note that the administration of GH to hypothyroid children with GH deficiency can increase hypothyroidism, something that does not occur when GH is administered once the hypothyroidism has been corrected.

2.6.2 Glucocorticoids and GH secretion

The effects of glucocorticoids on GH synthesis and secretion are complex and reflect the pluripotential nature of these hormones. On the one hand, glucocorticoids are essential for the maintenance of GH secretion; thus, patients with adrenocortical insufficiency present GH deficiency that can be corrected by substitutive treatment with these steroids [120]. However, excess glucocorticoids decrease GH secretion and longitudinal growth [121, 122]. Acute administration of glucocorticoids leads to a GH-secreting response that lasts for a few hours, although after this there is a total blockade of the release of the hormone [123, 124]. This is probably all due to the fact that glucocorticoids stimulate transcription of the GH gene [125] and increase the stability of its mRNA [126], as well as they increase the expression rate of the GHRH receptor in somatotrophs.

The negative effect of excess glucocorticoids would depend on its induction of SS release, mediated by an increase in the responsiveness of SS neurons to β-adrenergic stimulation, as it happens in the periphery (**Figure 10**). It may also be that glucocorticoids directly modulate SS synthesis, since there is a glucocorticoid response element in the SS gene.

2.6.3 Sex steroids and GH secretion

The existence of a pattern of sexual dimorphism in GH secretion, first described in rat [127], has been a clearly well-established concept for many years [128]. Sexual differentiation occurs during fetal life; sexually different fetal sex steroid hormones regulate striking differences in the number of GHRH and SS neurons, their responses to sex steroids once the puberty begins, the adult hypothalamic synapse and its organization, and perhaps the number of somatotrophs and their responsiveness [129]. In humans of both sexes, spontaneous GH secretion is low during childhood, and it is gradually increasing until just before puberty, when the release of the hormone increases strongly [130, 131] and sexual dimorphism begins to manifest. Therefore, it appears that sex steroids explain this maturational change in the pattern of GH secretion. The question is: where does this action of sex steroids take place?

After puberty, GH secretion is greater in women than in men, although this can be modified depending on the phase of the menstrual cycle; the late follicular phase

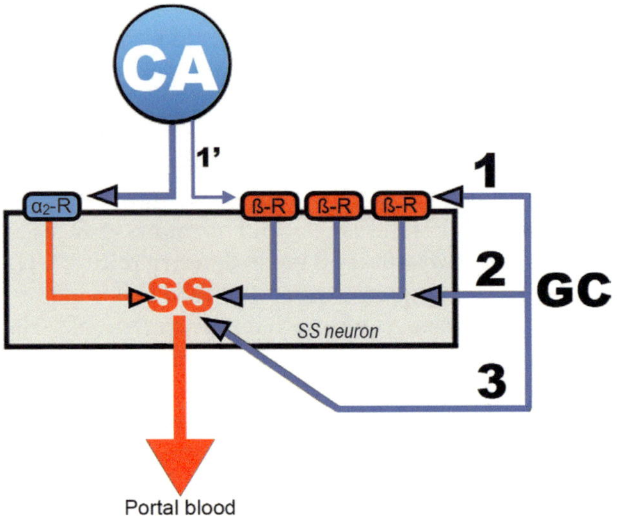

Figure 10.
Control of SS by glucocorticoids. Glucocorticoids (GC) stimulate the expression of β-adrenoceptors (β-R) in somatostatin neurons (SS neuron) and favor their response to low levels of catecholamines (1′ and 2), inducing the release of SS. Furthermore, GCs can stimulate the transcription of the SS gene (3). The result is an increased discharge of SS into the portal blood.

leads to GH secretion greater than that seen in both the early follicular phase and the luteal phase.

A direct pituitary effect of sex steroids on GH secretion cannot be excluded; however, they most likely act primarily on the hypothalamic level, modulating the rhythmic interaction of GHRH-SS and, in particular, the availability of SS [61]. The difference between hypothalamic levels of free estradiol (fE2) in man and women can explain the differences in the secretory pattern of SS and, therefore, why GH release is sexually dimorphic.

In women, there is a large variation in plasma E2 levels throughout the menstrual cycle. We will have to bear in mind that only fE2 is biologically active, and that only in this form can it cross the blood-brain barrier. In fact, most of E2, but also of T, circulates bound to sex hormone-binding globulin (SHBG), although its affinity for this carrier is lower than that of T. However, plasma levels of SHBG are approximately two times higher in women than in men, because E2 increases the hepatic synthesis of this globulin. This, together with the fact that T circulates in levels of ng/ml, while E2 does so in pg/ml, suggests that free testosterone (fT) reaches higher levels in the hypothalamus than fE2, but this would be compensated by the hypothalamic aromatization from T to E2.

The data in rats clearly indicate that E2 replacement therapy rapidly reverses the decrease in SS mRNA observed after oophorectomy [132], although a direct effect of E2 on the SS gene is unlikely. E2 can affect both the biosynthesis and turnover of catecholamines [133] in hypothalamic areas involved in the control of SS, as well as the responsiveness of α_2-adrenoceptors in SS neurons. This, together with our findings demonstrating the role of the adrenergic system in the regulation of SS [54, 61], allows to hypothesize that the action of E2 on SS neurons depends on its effects on adrenergic transmission to these neurons [4]. This is consistent with data indicating that, at the peripheral level, catecholamines appear to be integral signaling components for maintaining steroid sensitivity in some reproductive tissues [134].

In addition, E2 inhibits the hepatic synthesis of IGF-I, thus reducing the effects of this inhibitor of both SS and GH.

On the other hand, T seems to stimulate the release of SS, but it depends on its central aromatization to E2. Therefore, the different levels of free E2 fE2 related to gender could explain the dimorphism in GH secretion [4]. Indeed, in humans, plasma fE2 but not T, were shown to be strongly correlated with rates of total and pulsatile GH release [5]. Support for this approach is given by the fact that there is great aromatase activity in the hypothalamus, as well as by the fact that it is E2 and not T that modifies the activity of a series of enzymes involved in synthesis and turnover of catecholamines. This explains why in children with delayed puberty the administration of T produces an increase in the pituitary reserve of GH, which is not observed when a non-aromatizable androgen, such as oxandrolone, is administered. Furthermore, studies in pubertal boys and adult men given the antiestrogen tamoxifen abrogate the stimulatory effect of T on GH [135, 136].

In summary, higher levels of fE2 in the hypothalamus, in women, can cause decreased biosynthesis and/or release of SS. If fE2 exerts a positive effect on the production of catecholamines at the hypothalamic level, they will also act positively on the production of GHRH. In GHRH neurons, there is an estrogen receptor ERα whose deletion in mice delays female puberty, since in females a group of GHRH neurons changes their phenotype to start producing Kiss1 (a key peptide in gonadotropic regulation). Therefore, a direct action of estrogens on neurons with the dual GHRH/Kiss1 phenotype modulates growth and puberty and may direct the sex differences in endocrine function observed during pubertal transition [137].

In women, episodic GH secretion is more frequent, and plasma GH values during trough periods are somewhat higher than in men. Therefore, in general, the amount of GH secreted in peak episodes is higher in women. But this can change throughout the menstrual cycle, as described before. Pulsatile GH release in women is doubled in the late follicular stage [138], and when estrogens are administered for superovulation, GH release is significantly increased [139]. Moreover, treatment of men with diethylstilbestrol (a potent estrogen) induces a change in the pulsatile pattern of GH release similar to that commonly seen in females [140]. It is important to remark that in humans, as in the rat [141], sex steroids would act mainly during the fetal and early neonatal stages, impregnating the hypothalamus so that from puberty there is an increase in the pulsatile secretion of GH, determined by an increase in the amplitude of the peaks rather than a change in secretory frequency [142]. In adults, the modification of the levels of sex steroids is not accompanied by significant modifications of the secretion of the hormone [4, 135].

The fact that estrogens produce an alteration in GH secretion may be due to the affectation of the feedback mechanisms through which IGF-I acts on GHRH and SS, in this case extraordinarily diminished by these steroids [143]. Alteration in GH output in response to estrogens is thought to occur as a consequence of altered negative feedback, wherein the actions of IGF-I on hypothalamic components of the GH-axis are greatly diminished [143]. This is something paradoxical [144], with evidence suggesting a complex relationship between IGF-I, estrogens, and GH output [129]. At low doses, estrogens increase GH output (while IGF-I production is high), while high doses of estrogens can decrease bioactive IGF-I levels, preventing the inhibitory effect of IGF-I on GH secretion. Furthermore, the route by which estrogens are administered clearly affects their effects on GH secretion. The pulsatile secretion of GH is very different when E2 is administered orally than when it is administered *via* transdermal patches, in both premenopausal and postmenopausal women. In two studies, oral administration of estradiol caused an increase in total daily GH secretion and its pulsatile release and a decrease in plasma IGF-I levels, whereas transdermal estrogen treatment increased circulating levels of IGF-I in post-menopausal to that seen in pre-menopausal women without altering the pulsatile release of GH [145, 146]. Therefore, it is likely that oral administration of E2 leads to a decrease in plasma IGF-I levels, consequently reducing its inhibitory effects, at the pituitary and hypothalamic levels, on GH secretion. Therefore, it is likely that the decrease in GH secretion observed at menopause is not due to the reduction in estrogen levels [146], since, furthermore, neither oral nor transdermal administration of E2 in menopausal women is able to reverse the decreased GH secretion associated with aging even when stimulated with GHRH. Moreover, the route of estrogen administration significantly alters the effect of estradiol on GH release. Assessment of pulsatile GH output in pre- and postmenopausal women following oral versus transdermal estradiol treatment revealed divergent effects on GH release. In two studies, oral administration of estradiol caused an increase in total daily GH secretion and pulsatility and a decrease in plasma IGF-I levels. Increased 24 h mean and pulsatile GH release and decreased circulating levels of IGF-I, whereas transdermal estrogen treatment increased circulating levels of IGF-I in post-menopausal to that seen in pre-menopausal women without altering the pulsatile release of GH [145, 146]. Thus, it is thought that oral administration of estrogen increases GH release through lowering circulating levels of IGF-I, thereby reducing negative feedback. These observations further indicate that

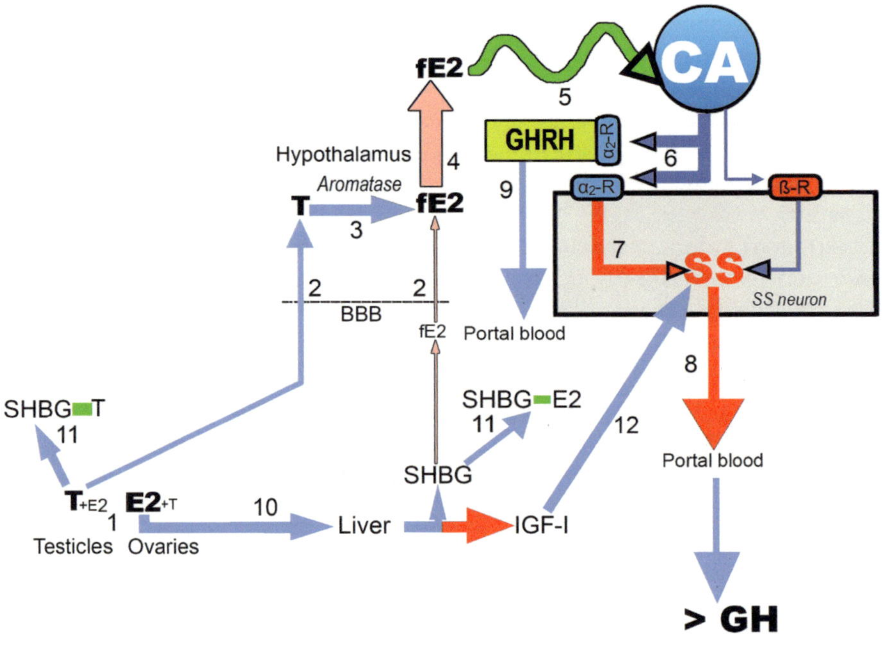

Figure 11.
Control of SS by sex steroids. Depending on the sex, the testicles or the ovaries secrete sex steroids in the blood (1), testosterone (T) or estradiol (E2), respectively, but also low amounts of E2 (testicles) and T (ovaries). While T can cross the blood-brain barrier (BBB, 2), only free E_2 (fE2, 2) can reach the brain. In the hypothalamus, T is aromatized to E2 (3), so the amount of fE2 is greater in women (4) than in men. fE2 modulates hypothalamic synthesis and catecholamine turnover (CA, 5), so the amount of catecholamines that reach SS neurons [5] is high enough to stimulate α2-adrenoceptors (6), inhibitors of the synthesis (7) and release of SS (8), but stimulators of GHRH synthesis (6) and release (9). E2 induces the hepatic synthesis of SHBG (10) and binds to it (11), as does T (11), but with lower affinity than T. Also in the liver, E2 decreases the synthesis of IGF-I (red arrow), a stimulator of SS release (12).

the reduced GH output following menopause is not due to a reduction in circulating levels estrogen [146]. To this extent, neither oral nor transdermal estrogen treatment in postmenopausal women can reverse age-related reductions in spontaneous or GHRH-stimulated GH and IGF-I release. However, a recent randomized, double-blind, controlled study, performed in 60 healthy postmenopausal women conclude that while there was a E2 decrease when giving anastrozole (an inhibitor of aromatase activity), but no changes in E2 levels were observed when fulvestrant (a selective estrogen receptor-α antagonist) was administered. In spite of it, GH rebound after SS infusion declined markedly during both estrogen-deprivation schedules, suggesting a previously unrecognized dependence of hypothalamic-pituitary GH regulation on low levels of endogenous estrogen during menopause [147].

As we have seen throughout this section, although it is clear that sex steroids play a more than important role in the control of GH secretion, with E2 as the main actor, there are many controversies regarding both the mechanism of action and the effects they produce. Most likely this is a reflection of the great complexity of the world of GH and the number of factors involved in it.

In a study from our group carried out in young healthy volunteers of both sexes, it was concluded that testosterone acts on the release of GH at the supra-hypophyseal level and that its action is mediated by its aromatization to E2 [148]. Therefore, the association between reduced GH release at menopause and low circulating estrogen levels must be thoroughly analyzed in order to understand the mechanisms that may link these concepts.

Another point of interest is the fact that in women there is no cortico-adrenal production of estrogens and there is production of androgenic precursors that give rise to testosterone in the liver, as in men, which, although small, can help in GH secretory control.

A summary of these concepts is shown in **Figure 11**.

2.6.4 Insulin and GH secretion

Insulin and GH play some opposite actions at the metabolic level in the human body. Therefore, it is logical to think that insulin has effects on pituitary GH secretion.

As is known, chronic GH secretion counteracts insulin effects leading to peripheral insulin insensitivity, resulting in pancreatic hypersecretion of insulin that attempts to compensate for its loss of sensitivity [149]. On the contrary, as already defined, weight gain leads to poor GH secretion correlated with an increase in circulating insulin. On these basis, it appears that GH release is tightly controlled by prolonged changes in energy intake and supposed insulin release. There are many situations that support these inverse relationships in various metabolic conditions. For example, fasting, anorexia nervosa, and type-1 diabetes are associated with decreased insulin levels and increased GH secretion, while hyperphagia and obesity are associated with increased insulin secretion and decreased GH secretion.

A series of many evidences points to a direct relationship between hypoinsulinemia and excess increased secretion of GH. Intensive insulin treatment in type-1 diabetic patients reverses GH hypersecretion [99], whereas calorie restriction to reverse hyperinsulinemia [150] results in the recovery of GH secretion, presumably in response to a reduction in insulin levels [151].

Insulin inhibits the expression of the GH gene in isolated pituitary cells [152, 153], as well as acts selectively on its receptors in the pituitary gland [154] in such a way that it suppresses GH release from isolated somatotrophs [154, 155]. These indicate that, although systemic insulin resistance exists [154], somatotrophs remain sensitive to insulin in obesity. Therefore, insulin can modulate GH release in relation to weight gain and promote continued suppression of GH release in obesity, independent of the inhibitory effect of plasma insulin. In contrast, *in vivo* deletion of somatotroph insulin receptors leads to increased pituitary GH content and release [155]. These effects occur independently of the development of systemic insulin resistance, suggesting that somatotrophs remain insulin sensitive in obesity. Thus, insulin can modulate GH release in relation to weight gain and promote sustained suppression of GH release in obesity, independent of systemic insulin resistance. In contrast, *in vivo* removal of insulin receptors from somatotrophs leads to an increase in the content and release of GH in the pituitary [155]. The suppressive actions of insulin on GH release occur independently of IGF-I, since the selective inactivation of the insulin receptor and IGF-I receptor (Insr and Igflr) genes in mouse somatotrophs does not affect these effects of insulin [156], despite the fact that in the liver, insulin facilitates the expression of IGF-I, which suppresses the synthesis and secretion of GH. Therefore, it is likely that insulin can provide critical feedback to alter GH release in relation to long-term metabolic requirements. It should be noted that loss of somatotroph-specific insulin receptor expression does not completely reverse the suppression of GH release seen during diet-induced weight gain and thus leads to obesity [156].

The actions of insulin on GH release may not be restricted to the pituitary. Insulin receptors are expressed throughout the hypothalamus, including the arcuate and periventricular nuclei [157–160]; therefore, insulin can modulate GH release by acting through the hypothalamus, although this is not yet well known. As insulin does not predict ultradian GH release patterns, the actions of insulin at the hypothalamus level to mediate GH production may be limited to infradian regulation of maximal GH production [130].

2.6.5 IGF-I and GH secretion

IGF-I is the main mediator of the actions of GH in the body and shares many effects with it, although it also has specific effects of its own. IGF-I is produced in virtually any tissue, but plasma IGF-I is synthesized and released from the liver. There are specific relationships between IGF-I and GH, since IGF-I is part of a long feedback loop by which it inhibits GH secretion, acting both at the hypothalamic level, where it stimulates the release of SS, and at the pituitary level, inhibiting GH secretion directly. Even at the level of the median eminence, specific binding sites for IGF-I have been identified where it acts by inhibiting GHRH secretion [161, 162]. At the pituitary level, IGF-I inhibits the transcription of GH and Pit-1 genes, both under basal conditions and after stimulation with GHRH [41, 163].

2.6.6 Ghrelin, Klotho, and GH secretion

At the end of the past century, Bowers' group [164] tried to identify how peptides derived from endogenous opioids could stimulate GH secretion. They synthesized a Met-enkephalin derivative, without opioid activity, which they named GHRP-6 (growth hormone-releasing peptide-6) [165]; it exhibited GH-releasing activity *in vivo,* in humans, acting synergistically with GHRH without binding to any known receptor. After the synthesis of this GH secretagogue, many other were synthesized, until a specific receptor for these synthetic compounds was identified in 1996 [166]. This receptor was a GTP-binding protein (GHSR-1a) present in the pituitary and arcuate and infundibular hypothalamus of several species, including humans [166], and its detection implied that a natural hormone had to exist. Indeed, this hormone was found in the stomach of rats and was called ghrelin [167].

Since we recently published an extensive review of the actions of ghrelin and klotho [168], only a few of the many important actions of these peptides will be discussed in this chapter.

2.6.6.1 Ghrelin and GH secretion

Ghrelin is a 28 amino acid peptide, which is in fact a true pleiotropic hormone. In spite of its peptide structure, ghrelin must bind to an octanoyl group on the third amino acid to acquire full biological activities [169].

The discovery of ghrelin and the fact that it acts synergistically with GHRH caused a change in the concept of basic regulation of GH; that is, in addition to the classical GHRH-SS interaction, GH secretion from the pituitary gland is also modulated by ghrelin produced by specific cells located in oxyntic cells in the stomach (**Figure 12A**). Ghrelin is an orexigenic hormone that appears in the blood during fasting and reaches the CNS to transmit a hunger signal. This explains why in anorexia nervosa there is an increase in plasma ghrelin concentrations, while in obese subjects it is reduced, and also why GH secretion is increased in anorexia nervosa patients and during fasting, while it is reduced or absent in obesity, but also in the elderly [170] (**Figure 12A**). Therefore, it seems that ghrelin appeared in evolution to stimulate eating behavior and optimize the use of digested food by promoting the secretion of an anabolic hormone, such as GH [171].

As indicated above, the active form of ghrelin that induces GH secretion, by stimulating GHRH secretion and inhibiting SS and IGF-I secretion (**Figure 12A**),

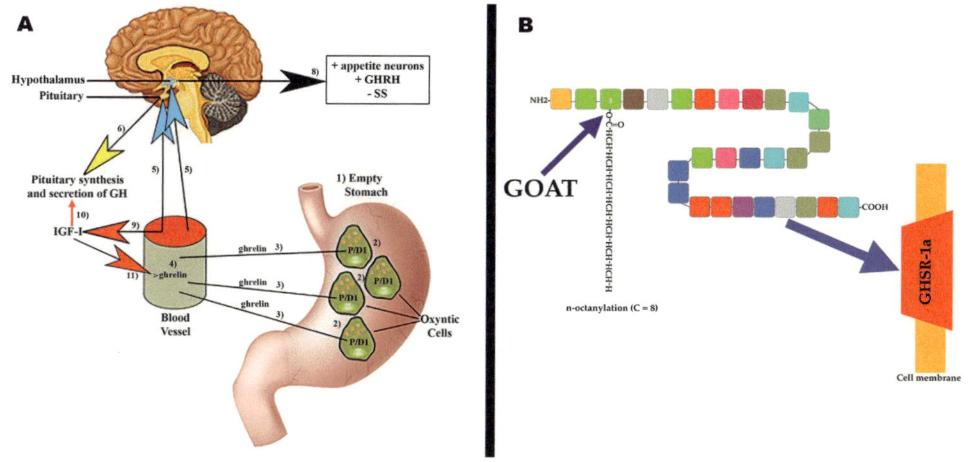

Figure 12.
Control of GH secretion by ghrelin and GH secretion. A: In fasting situations, the empty stomach (1) produces ghrelin in P/DI cells (2), which is released to the blood (3, 4) from where it reaches the pituitary gland (5) and induces the synthesis and secretion of GH (6). Ghrelin also reaches the hypothalamus (7), where it stimulates (8) appetite neurons and GHRH and inhibits SS. Plasma ghrelin also inhibits IGF-I (9), therefore inhibiting its inhibitory effects on the pituitary synthesis and secretion of GH (10). In turn, IGF-I inhibits the increase in circulating ghrelin (11) produced by fasting. Blue arrows: stimulation. Red arrows: inhibition. Yellow arrow: synthesis and secretion of GH. Black arrow: hypothalamic effects of ghrelin. +: stimulation. –: inhibition. B: The active form of ghrelin is acyl ghrelin, obtained after an acylation performed by GOAT, which links after linking a fatty acid side chain to serine 3. Acyl ghrelin can then bind to its receptor GHSR-1a and activate it. This figure has been modified from reference 169, Front Endocrinol. 2021.

and many other physiological functions, such as inhibiting cell death in cardiac and endothelial cells [172], is the acylated form (acyl ghrelin), produced by attaching an octanoyl group to amino acid 3 (serine) carried out by ghrelin O-acyl transferase (GOAT) (**Figure 12B**) [173, 174].

Ghrelin stimulates GH release: (1) directly on the pituitary somatotrophs and (2) antagonizing SS and inducing GHRH secretion. In addition, ghrelin decreases plasma levels of IGF-I, therefore inhibiting the negative effect of IGF-I on GH secretion.

In humans, the physiological nocturnal rise in GH was not inhibited by the infusion of a powerful analog of SS, such as octreotide [175, 176]. This indicates that the effect of ghrelin was not affected by SS. On the other hand, patients with inactive GHRH receptor, due to mutations, still had rhythmic GH secretion, suggesting that another factor, different to GHRH, was inducing pituitary GH secretion [177]. However, a non-sense mutation affecting ghrelin receptor in humans was found to be associated with short stature [178]. Furthermore, in rats, intravenous (iv) administration of ghrelin during a GH peak induced a marked increase in plasma GH [179], but the immunoneutralization of GHRH led to a virtual absence of ghrelin-induced GH secretion; however, when ghrelin was administered during a physiological trough period, the GH response was clearly diminished. Both effects indicate that ghrelin acts synergistically with GHRH while inhibits hypothalamic SS. In fact, ghrelin expression has been found in the hypothalamic arcuate nucleus [180].

Ghrelin and its receptor are also expressed in the pituitary [181, 182], so pituitary ghrelin may play an auto/paracrine role in the regulation of GH release. GHRH infusion increases pituitary ghrelin mRNA levels, suggesting that GHRH may be a regulator of pituitary ghrelin production [183]. Therefore, pituitary ghrelin can act physiologically on GH secretion, enhancing the response of somatotrophs to GHRH.

Ghrelin, as GHRH, acts through a G-protein-coupled receptor (GPCR), but in this case the activation of this receptor leads to the stimulation of the activity of phospholipase C (PLC) that induces the formation of IP3 and DAG (**Figure 13**); both IP3 and DAG induce an increase in cytosolic Ca^{2+} that facilitates the release of GH [179]. Ghrelin requires activation of the NOS/NO pathway and its subsequent guanylate cyclase (GC)/cGMP signal transduction pathway to induce GH release from the pituitary [85]. Chronic treatment with ghrelin produced an upregulation of GH transcription levels, as well as that of two isoforms of Na^+ channels, sensitive to blockade with tetrodotoxin (TTX), expressed in somatotrophs, such as NaV1.1 and NaV1.2. This indicates that ghrelin also regulates the expression of the Na^+ channel gene in somatotrophs (**Figure 13**) [184].

Interestingly, the hypothalamic enzymatic complex AMP-activated protein kinase (AMPK), involved in the hypothalamic control of energy and metabolic homeostasis, is also activated by ghrelin. Therefore, AMPK also participates in the control of GH secretion, and as its blockade or its functional impairment inhibits ghrelin- or GHRH-induced GH secretion, most likely by increasing SS production tone [185] (**Figure 13**).

It is well known that aging is associated with a decrease in GH secretion from the second decade of life [186]. Something similar happens with ghrelin [170, 187], although the pituitary ghrelin receptor does not decline with aging and the GH response to ghrelin is still seen in the elderly and there is an age-related decline [188].

Therefore, what is the reason why the secretion of an orexigenic hormone, such as ghrelin, and also that of an anabolic hormone, such as GH, is lost with aging?

Gastric ghrelin synthesis and secretion increase during fasting and decrease during feeding [189]. This is the reason why chronic intake of high-calorie diets, prolonged ingestion of high fats, and obesity lead to a reduction in gastric ghrelin

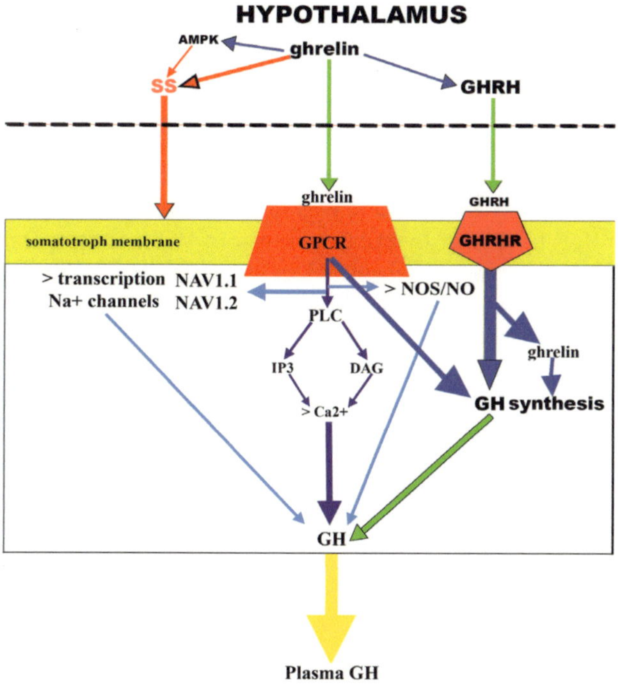

Figure 13.
Ghrelin control of GH secretion. Mechanisms of action of ghrelin in the control of GH secretion. Hypothalamic ghrelin (coming from the circulation and/or synthetized in the hypothalamus) stimulates the release of GHRH into the portal blood from where GHRH reaches the somatotrophs (green arrow) and induces synthesis (blue arrow) and release (green arrow) of GH, but also that of pituitary ghrelin that also participates in the synthesis of GH (blue arrow). Hypothalamic ghrelin antagonizes SS by itself (red arrow) but also through a mechanism mediated by its stimulation of AMPK (blue arrow). At the pituitary level, ghrelin from the hypothalamus (green arrow) or from the systemic circulation stimulates the release of GH via activation of the PLC that activates the intracellular Ca^{2+} supply and the transcription of Na^{+} channels necessary for the release of GH. Likewise, ghrelin after the activation of its membrane receptor (GPCR) activates the NOS/NO pathway, also necessary for the release of GH. Taken from reference [169], Front Endocrinol. 2021.

production and secretion [189, 190], while a low protein supply significantly increases plasma ghrelin [190].

Adrenergic hormones stimulate the release of gastric ghrelin by acting directly on the β1 receptors of ghrelin-producing cells, very rich in this type of adrenergic receptors [191]. Therefore, fasting acts on gastric ghrelin secreting cells through the sympathetic nervous system, something that appears to be logical given the relationships between this autonomic system and the hypothalamic production of GHRH and SS for controlling GH secretion. The administration of muscarinic agonists also increases plasma ghrelin concentrations [192], as does vagus nerve excitation in the gastric mucosa [193].

Furthermore, and given the relationships between ghrelin, SS, and GH secretion, it stands to reason that SS [194] and GH [195] inhibit gastric ghrelin secretion, although ghrelin also stimulates pancreatic SS production [196], to inhibit insulin release [197, 198] and contribute to glucose homeostasis [197].

Since, as we have seen, metabolic factors play an important role in the control of GH secretion, it is logical that they also participate, *via* hormonal modulation, in the control of gastric ghrelin secretion, as can be seen in detail in Ref. [169]. Perhaps the only hormones to be discussed in this section are leptin and IGF-I. Leptin, a hormone produced by adipocytes, which acts by decreasing food intake; that is, it is an anorexigenic hormone, and it is logical that it acts by decreasing plasma ghrelin and inhibiting its actions at the central level [199].

Similarly, logical are the relationships between IGF-I and ghrelin. Plasma IGF-I concentration significantly determines plasma ghrelin concentrations, existing a negative correlation between them [200]. This has been found in children and adolescents [201, 202]. On this basis, the highest concentration of ghrelin was observed in GH-deficient children in whom there was low availability of free IGF-I [203], the bioactive form of IGF-I. That is, low plasma levels of IGF-I induce the synthesis and secretion of ghrelin, while in turn ghrelin decreases plasma levels of IGF-I.

2.6.6.2 Klotho and GH secretion

Although klotho was identified in 1997 as an antiaging agent, klotho-deficient mice exhibit growth retardation [204], implying that this transmembrane protein is also involved in the control of GH secretion. The kidneys are the main source of klotho, where it is stimulated by insulin [205] and IGF-I [206]. In contrast, klotho inhibits both receptor activation and intracellular signaling of these hormones [207], thus acting as a positive regulator of GH secretion.

Klotho induces GH secretion by activating the extracellular signal regulated kinase 1/2 (ERK1/2) pathway in AP somatotrophs [208]. Relationships between klotho and GH can be seen in untreated GH-deficient children and adults; in these patients, the plasmatic levels of klotho are low, while treatment with GH normalizes this klotho deficit, through a mechanism mediated by the activation of the Akt-mTOR pathway [206], a key pathway in GH signaling.

In addition to its production in the kidney, klotho is also produced in the somatotrophs, most likely to modulate auto/paracrine GH secretion. Interestingly, high production of klotho has been observed in pituitary adenomas that do not secrete GH, suggesting that klotho is also produced in other pituitary cells which are also a source of this still little known hormone that plays so many different roles in the body.

2.6.7 Neuropeptide Y (NPY) and GH secretion

NPY is a 36 amino-acid peptide belonging to the family of pancreatic polypeptides. It is expressed in the hypothalamus in neurons of the arcuate, paraventricular, and periventricular nuclei [209]. NPY exerts an important effect on the regulation of energy intake and expenditure, something that, as we have already analyzed, once again involves GH, so NPY has to play a role in regulating the secretion of this hormone, although this role is still little known. Ancient studies carried out in rodents showed that NPY inhibited the GH axis [210], most likely due to the fact that NPY neurons in the arcuate nucleus connect with periventricular SS neurons [211]. In fact, after 9 hours of fasting, it was observed, in mice, that the hypothalamic levels of NPY increased as did those of SS [95], which suggested that NPY would be the factor responsible for the inhibition of GH secretion, mediated by SS, associated with fasting. There are five receptors for NPY. Among them the Y1 receptor is mainly expressed at the post-synaptic level, while the Y2 receptor is expressed in NPY neurons [212, 213].

A study carried out in mice indicates that activation of Y1 receptor inhibits the secretion of GH during fasting, probably through induction of SS release, whereas Y2 receptor has no effect on GH secretion in fasting conditions, but maintains pulsatile GH release in mice fed *"ad libitum"* [96].

All these data indicate that NPY neurons play a very important role in the control of energy intake and expenditure, as indicated above. This concept is reinforced by the fact that most of these neurons co-express NPY and Agouti-related peptide (AgRP, another important regulator of metabolic homeostasis), and more importantly, there is an insulin/NPY network that regulates energy homeostasis, so that the lack of insulin signaling in NPY neurons induces an increase in energy reserves and obesity and also produces an alteration of GH/IGF-I axis. This is a clear example of how GH control is key to proper homeostasis of metabolism and food consumption, and how these, in turn, control GH production.

3. Facts and perspectives

It is now clear that pituitary secretion of GH is mainly subject to a tonic inhibitory control exerted by SS, contrary to what was thought after the discovery of GHRH. Therefore, only the factors that can inhibit the production of SS will be responsible for the release of GH induced by GHRH, whose synthesis and release into the portal circulation are inhibited by SS. All this was identified in the last decade of the last century. However, the great complexity of the hypothalamus, in terms of the large number of existing neuronal types that can release very different compounds, suggests that more neurotransmitters with the capacity to act on the SS and, consequently, on the hypothalamic-somatotropic axis can still be identified, although, logically, their actions will always be more secondary than those of the factors that have been described in this chapter.

4. Conclusions

In this chapter, the main factors involved in the control of GH secretion were analyzed, including the explanation of why it is sexually dimorphic or its relationship with feeding or fasting.

Although GH has traditionally been considered a merely metabolic hormone with specific actions on the longitudinal growth of the organism until the end of puberty, a series of studies carried out in recent years show that this hormone is highly complex in terms of the number of actions so diverse that it exerts in the organism. This has led to the control of its secretion being very diverse, although it primarily depends on a control exerted negatively from the hypothalamus by somatostatin and an inducing effect of its release carried out by GHRH, although it also depends on the nutritional status, by a primarily gastric peptide such as ghrelin.

As is logical, given the multiplicity of GH actions, a series of hormonal, metabolic factors, or factors simply indicative of the functional state of a gland or tissue, or neurotransmitters, will condition the rate and type of GH secretion. Basically, these factors will act primarily on the secretion of somatostatin, activating it and, consequently, inhibiting the secretion of GHRH-stimulating peptide, or inhibiting SS so that GHRH can be released into the hypothalamic-pituitary portal circulation, and stimulate the synthesis and pituitary secretion of GH. The basic control of SS secretion depends on the adrenergic pathways that reach the neurons that produce it, being the α_2-adrenergic inhibitory and the β-stimulants of its secretion, although some other factors, such as arginine or ghrelin, may act directly on SS neurons, while others, such as thyroid hormones, may act by regulating the number and responsiveness of GHRH receptors in the pituitary gland.

Conflict of interest

The authors declare no conflict of interest.

Author details

Jesús Devesa[1*] and Pablo Devesa[2]

1 Scientific Direction, Medical Center Foltra, Teo, Spain

2 Coordination and Research, School of Medicine, University Univalle, Cochabamba, Bolivia

*Address all correspondence to: jesus.devesa@usc.es

References

[1] Robinson ICAF, Clark RG. The secretory pattern of GH and its significance for growth in the rat. In: Isaksson O, Binder C, Hall K, Hökfelt B, editors. Growth Hormone: Basic and Clinical Aspects. Amsterdam: Elsevier; 1987. pp. 109-127

[2] Tapanainen P, Ranala H, Leppäluoto J, Lautala P, Käär ML, Knip M. Nocturnal release of immunoreactive growth hormone-releasing hormone and growth hormone in normal children. Pediatric Research. 1989;**26**:404-409

[3] Brandenberger G, Weibel L. The 24-h growth hormone rhythm in men: Sleep and circadian influences questioned. Journal of Sleep Research. 2004;**13**:251-255

[4] Devesa J, Lois N, Arce V, Díaz MJ, Lima L, Tresguerres JAF. The role of sexual steroids in the modulation of growth hormone secretion in humans. The Journal of Steroid Biochemistry and Molecular Biology. 1991;**40**:165-173

[5] Ho KY, Evans WS, Blizzard RM, Veldhuis JD, Merriam GR, Samojlik E, et al. Effects of sex and age on the 24-hour profile of growth hormone secretion in man: Importance of endogenous estradiol concentrations. The Journal of Clinical Endocrinology and Metabolism. 1987;**64**:51-58

[6] Jansson JO, Frohman LA. Inhibitory effect of the ovaries on neonatal imprinting of growth hormone secretion in female rats. Endocrinology. 1987;**121**:1417-1423

[7] Plotsky P, Vale W. Patterns of growth hormone-releasing hormone and somatostatin secretion into the hypophyseal-portal circulation. Science. 1985;**230**:461-463

[8] Devesa J, Lima L, Lois N, Fraga C, Lechuga MJ, Arce V, et al. Reasons for the variability in growth hormone [GH] responses to GHRH challenge: The endogenous hypothalamic-somatotroph rhythm [HSR]. Clinical Endocrinology. 1989;**30**:367-377

[9] Reichlin S. Growth and the hypothalamus. Endocrinology. 1960;**67**:760-773

[10] Rivier J, Spiess J, Thorner MO, Vale W. Characterization of a growth hormonereleasing factor from a human pancreatic islet tumor. Nature. 1982;**300**:276-278

[11] Guillemin R, Brazeau P, Bohlen P, Esch F, Ling N, Wehrenberg WB. Growth hormone releasing factor from a human pancreatic tumor that caused acromegaly. Science. 1982;**218**:585-587

[12] Frohman LA, Szabo M. Ectopic production of growth hormone-releasing hormone by carcinoid and pancreatic islet tumors associated with acromegaly. In: Physiopathology of Endocrine Diseases and Mechanisms of Hormone Action. New York: Liss; 1981. pp. 259-271

[13] Cronin MJ, Rogol AD, Dabney LG, Thorner MO. Selective growth hormone and cyclic AMP-stimulating activity is present in a human pancreatic islet cell tumor. The Journal of Clinical Endocrinology and Metabolism. 1982;**55**:381-383

[14] Mayo KE, Cerelli GM, Lebo RV, Bruce BD, Rosenfeld MG, Evans RM. Gene encoding human growth hormone-releasing factor precursor: Structure sequence and chromosomal assignment. Proceedings of the National Academy of

Sciences of the United States of America. 1985;**82**:63

[15] Mutsuga N, Iwasaki Y, Morishita M, Nomura A, Yamamori E, Yoshida M, et al. Homeobox protein Gsh-1-dependent regulator of the rat GHRH gene promoter. Molecular Endocrinology. 2001;**15**:2149-2156

[16] Mantamadiotis T, Kretz O, Ridder S, Bleckmann SC, Bock D, Gröne H-J, et al. Hypothalamic 3′,5′ cyclic adenosine monophosphate response element-binding protein loss causes anterior pituitary hypoplasia and dwarfism in mice. Molecular Endocrinology. 2006;**20**:204-211

[17] Asai M, Iwasaki Y, Yoshida M, Mutsuga-Nakayama N, Arima H, Ito M, et al. Nuclear factor of activated T cells [NFAT] is involved in the depolarization-induced activation of growth hormone-releasing hormone gene transcription in vitro. Molecular Endocrinology. 2004;**18**:3011-3019

[18] Brazeau P, Vale W, Burgus R, Ling N, Butcher M, Rivier J, et al. Hypothalamic polypeptide that inhibits the secretion of immunoreactive pituitary growth hormone. Science. 1973;**179**:77-79

[19] Siler TM, Van den Berg G, Yen SS, Brazeau P, Vale W, Guillemin R. Inhibition of growth hormone release in humans by somatostatin. The Journal of Clinical Endocrinology and Metabolism. 1973;**37**:632-634

[20] Ling N, Burgus R, Rivier J, Vale W, Brazeau P. The use of mass spectrometry in deducing the sequence of somatostatin—A hypothalamic polypeptide that inhibits the secretion of growth hormone. Biochemical and Biophysical Research Communications. 1973;**50**:127-133

[21] Naylor S, Sakaguichi A, Shen LP, Bell GI, Rutter WJ, Shows TB. Polymorphic human somatostatin gene is located on chromosome 3. Proceedings of the National Academy of Sciences of the United States of America. 1983;**80**:2686

[22] Shen LP, Rutter WJ. Sequence of the somatostatin I gene. Science. 1984;**224**:168

[23] Goodman RH. Regulation of neuropeptide gene expression. Annual Review of Neuroscience (Palo Alto, CA). 1990;**13**:111

[24] Pradayrol L, Jornvall H, Mutt V, Ribet A. N-terminally extended somatostatin: The primary structure of somatostatin-28. FEBS Letters. 1980;**109**:55

[25] Rai U, Thrimawithana TR, Valery C, Young SA. Therapeutic uses of somatostatin and its analogues: Current view and potential applications. Pharmacology & Therapeutics. 2015;**152**:98-110

[26] Mayo KE. Molecular cloning and expression of a pituitary-specific receptor for growth hormone-releasing hormone. Molecular Endocrinology. 1992;**6**:1734

[27] Mayo KE, Godfrey PA, Suhr ST, Kulik DJ, Rahal JO. Growth hormone-releasing hormone: Synthesis and signaling. Recent Progress in Hormone Research. 1995;**50**:35

[28] Vamvakopoulos NC, Kunz J, Olberding U, Scherer SW, Sioutopoulou TO, Schneider V, et al. Mapping the human growth hormone-releasing hormone receptor [GHRHR] gene to the short arm of chromosome 7 [7p13-p21] near the epidermal growth factor receptor [EGFR]. Genomics. 1994;**20**:338-340

[29] Spengler D, Waeber C, Pantaloni C, Holsboer F, Bockhaert J, Seeburg PH,

et al. Differential signal transduction by five splicing variants of the PACAP receptor. Nature. 1993;**365**:170-175

[30] Lin C, Lin SC, Chang CP, Rosenfeld MG. Pit-1-dependent expression of the receptor for growth hormone releasing factor mediates pituitary cell growth. Nature. 1992;**360**:765-768

[31] Gaylinn BD, Harrison JK, Zysk JR, Lyons CE, Lynch KR, Thorner MO. Molecular cloning and expression of a human anterior pituitary receptor for growth hormone-releasing hormone. Molecular Endocrinology. 1993;**7**:77-84

[32] Gonzalez GA, Montminy MR. Cyclic AMP stimulates somatostatin gene transcription by phosphorylation of CREB at serine 133. Cell. 1989;**59**:675-680

[33] Sheng M, Thompson MA, Greenberg ME. CREB: A Ca[2+]-regulated transcription factor phosphorylated by calmodulin-dependent kinases. Science. 1991;**252**:1427-1430

[34] Ingraham HA, Chen RP, Mangalam HJ, Elshotz HP, Flynn SE, Lin CR, et al. A tissue-specific transcription factor containing a homeodomain specifies a pituitary phenotype. Cell. 1988;**55**:519-529

[35] McCormick A, Brady H, Theill LE, Karin M. Regulation of the pituitary-specific homeobox gene GHF1 by cell-autonomous and environmental cues. Nature. 1990;**345**:829-832

[36] Wang B, Qin C, Zhang C, Jia J, Sun C, Li W. Differential involvement of signaling pathways in the regulation of growth hormone release by somatostatin and growth hormone-releasing hormone in orange-spotted grouper [Epinephelus coioides]. Molecular and Cellular Endocrinology. 2014;**382**:851-859

[37] Zeitler P, Siriwardana G. Stimulation of mitogen-activated protein kinase pathway in rat somatotrophs by growth hormone releasing hormone. Endocrine. 2020;**12**:257-264

[38] Baldwin JM. Structure and functions of receptors coupled to G-proteins. Current Opinion in Cell Biology. 1994;**6**:180

[39] Reisine T, Bell GI. Molecular biology of somatostatin receptors. Endocrine Reviews. 1995;**16**:427

[40] Raynor K, Murphy WA, Coy DH, Taylor JE, Moreau JP, Yasuda K, et al. Cloned somatostatin receptors: Identification of identification of subtype-selective peptides and demonstration of high affinity binding of linear peptides. Molecular Pharmacology. 1993;**43**:838-844

[41] Bertherat J, Bluet-Pajot MT, Epelbaum J. Neuroendocrine regulation of growth hormone. European Journal of Endocrinology. 1995;**132**:12-24

[42] Theill LE, Karin M. Transcriptional control of GH expression and anterior pituitary development. Endocrine Reviews. 1993;**14**:670-689

[43] Buscail L, Delesque N, Estève JP, Saint-Laurent N, Prats H, Clerc P, et al. Stimulation of tyrosine-phosphatase and inhibition of cell proliferation by somatostatin analogues: Mediation by human somatostatin receptor subtypes SSTR-1 and SSTR-2. Proceedings of the National Academy of Sciences of the United States of America. 1994;**91**:2315-2319

[44] Buscail L, Esteve JP, Saint-Laurent N, Bertrand V, Reisine T, O'Carroll AM, et al. Inhibition of cell proliferation by the somatostatin analogue RC-160 is mediated by SSTR-2 and SSTR-5 somatostatin receptors through different mechanisms. Proceedings of the National

Academy of Sciences of the United States of America. 1995;**92**:1580-1584

[45] Ruvkun G, Finney M. Regulation of transcription and cell identity of by POU domain proteins. Cell. 1991;**64**:475

[46] Ohta K, Nobukuni Y, Mitsubuchi H, Fujimoto S, Matsuo N, Inagaki H, et al. Mutations in the Pit-1 gene in children with combined pituitary hormone deficiency. Biochemical and Biophysical Research Communications. 1992;**189**:851-855

[47] Pfäffe RW, DiMattia GE, Parks JS, Brown MR, Wit JM, et al. Mutation of the specific POU domain of Pit-1 and hypopituitarism without pituitary hypoplasia. Science. 1992;**257**:1118-1121

[48] Delhase M, Bergani P, Malur A, Velkeniers B, Teugels E, Trouillas J, et al. Pit-1/GHF-1 expression in pituitary adenomas: Further analogy between human adenomas and rat SNtTW tumors. Journal of Molecular Endocrinology. 1993;**11**:129-139

[49] Delhase M, Rajas F, Verdood P, Remy C, Chevallier P, Velkeniers B, et al. Prolactin and growth hormone mRNA and protein characterization in SNtTW rat pituitary tumors. Journal of Molecular Endocrinology. 1995;**15**:233-243

[50] Delhase M, Castrillo JL, De la Hoya M, Rajas F, Hoogehe-Peters EL. AP-1 and Oct-1 transcription factors down-regulate the expression of the human Pit-1/GHF-1 gene. The Journal of Biological Chemistry. 1996;**271**:32349-32358

[51] Theill IE, Hattori K, Lazzaro D, Castrillo JM, Karin M. Differential splicing of the GHF-1 primary transcript gives rise to two functional distinct homeodomain proteins. The EMBO Journal. 1992;**11**:2261

[52] Stewart JK, Koerker DJ, Goodner CJ, Gale CC, Minton MF, Steiner RA. Regulation by catecholamines of spontaneous growth hormone secretion in the baboon. The American Journal of Physiology. 1981;**241**(3):E196-E199

[53] Durand D, Martin JB, Brazeau P. Evidence for a role of alpha-adrenergic mechanisms in regulation of episodic growth hormone secretion in the rat. Endocrinology. 1977;**100**:722-728

[54] Müller E. Neural control of somatotropic function. Physiological Reviews. 1987;**67**:962-1053

[55] Devesa J, Arce V, Lois N, Tresguerres JAF, Lima L. α_2-Adrenergic agonism enhances the growth hormone [GH] response to GH-releasing hormone through an inhibition of hypothalamic somatostatin release in normal men. The Journal of Clinical Endocrinology and Metabolism. 1990;**71**:1581-1588

[56] Arce V, Cella SG, Loche E, Ghigo E, Devesa J, Müller EE. Synergistic effect of growth hormone-releasing hormone [GHRH] and clonidine in stimulating GH release in young and old dogs. Brain Research. 1990;**537**:359-362

[57] Lima L, Arce V, Tresguerres JA, Devesa J. Clonidine potentiates the growth hormone [GH] response to GH-releasing hormone in norepinephrine synthesis-inhibited rats: Evidence for an alpha-2-adrenergic control of hypothalamic release of somatostatin. Neuroendocrinology. 1993;**57**:1155-1160

[58] Minamitani N, Chihara K, Kaji H, Kodama H, Kita T, Fujita T. α_2-Adrenergic control of growth hormone [GH] secretion in conscious male rabbits: Involvement of endogenous GH-releasing factor and somatostatin. Endocrinology. 1989;**125**:2839-2845

[59] Kelijman M, Frohman LA. ß-Adrenergic modulation of growth hormone [GH] auto feedback on sleep-associated and pharmacologically induced GH secretion. The Journal of Clinical Endocrinology and Metabolism. 1989;**69**:1187-1194

[60] Martha PM Jr, Blizzard RM, Thorner MO, Rogol AD. Atenolol enhances nocturnal growth hormone [GH] release in GHdeficient children during long term GH-releasing hormone therapy. The Journal of Clinical Endocrinology and Metabolism. 1990;**70**:56-61

[61] Devesa J, Díaz MJ, Tresguerres JAF, Arce V, Lima L. Evidence that α_2-adrenergic pathways play a major role in growth hormone [GH] neuroregulation: α_2-adrenergic agonism counteracts the inhibitory effect of muscarinic cholinergic receptor blockade on the GH response to GH-releasing hormone, while a2-adrenergic blockade diminishes the potentiating effect of increased cholinergic tone on such stimulation in normal men. The Journal of Clinical Endocrinology and Metabolism. 1991;**73**:251-256

[62] Richardson SB, Twente S. Inhibition of hypothalamic somatostatin release by beta-adrenergic antagonist. Endocrinology. 1990;**126**:1043-1046

[63] Vance ML, Kaiser D, Frohman LA, Rivier J, Vale W, Thorner MO. Role of dopamine in the regulation of GH secretion: Dopamine and bromocriptine augment GHRH-stimulated GH secretion in normal man. The Journal of Clinical Endocrinology and Metabolism. 1987;**64**:1136-1141

[64] Miell JP, Corder R, Pralong FP, Gaillard RC. Effects of dexamethasone on GHRH, arginine and dopaminergic stimulated GH secretion, and total plasma insulin-like growth factor-I concentrations in normal male volunteers. The Journal of Clinical Endocrinology and Metabolism. 1991;**72**:675-681

[65] Arce V, Lima L, Lois N, Rodríguez A, Díaz MJ, Tresguerres JA, et al. Role of central dopaminergic pathways in the neural control of growth hormone secretion in normal men: Studies with metoclopramide. Neuroendocrinology. 1991;**53**:143-149

[66] Bruni JF, Meites J. Effects of cholinergic drugs on growth hormone release. Life Sciences. 1978;**23**:1351-1357

[67] CasanuevaFF,VillanuevaL,CabranesJA, Cabezas-Cerrato J, Fernandez-Cruz A. Cholinergic mediation of growth hormone secretion elicited by arginine, clonidine, and physical exercise in man. The Journal of Clinical Endocrinology and Metabolism. 1984;**59**:526-530

[68] Locatelli V, Torsello A, Redaelli M, Ghigo E, Massare F, Muller EE. Cholinergic agonist and antagonist drugs modulate the growth hormone response to growth hormone-releasing hormone in the rat: Evidence for mediation by somatostatin. The Journal of Endocrinology. 1986;**111**:271-278

[69] Martin JB, Durand D, Gurd W, Faille G, Audet J, Brazeau P. Neuropharmacological regulation of episodic growth hormone and prolactin secretion in the rat. Endocrinology. 1978;**102**:106-113

[70] Wehrenberg WB, Wiviott SD, Voltz DM, Giustina A. Pyridostigmine-mediated growth hormone release: Evidence for somatostatin involvement. Endocrinology. 1992;**130**:1445-1450

[71] Mota A, Bento A, Peñalva A, Pombo M, Diéguez C. Role of serotonin receptor subtype 5-HT1D on basal and stimulated growth hormone secretion.

The Journal of Clinical Endocrinology and Metabolism. 1995;**80**:1973

[72] Murakami Y, Kato Y, Kabayama Y, Tojo K, Inoue T, Imura H. Involvement of growth hormone [GH]-releasing factor in GH secretion induced by serotoninergic mechanisms in conscious rats. Endocrinology. 1986;**119**:1089-1092

[73] Smythe GA, Lazarus L. Suppression of human growth hormone secretion by melatonin and cyproheptadine. The Journal of Clinical Investigation. 1974;**54**:116-121

[74] McCann SM, Vijayan E, Negro-Vilar A, Mizunuma H, Mangat H. Gamma aminobutyric acid [GABA], a modulator of anterior pituitary hormone secretion by hypothalamic and pituitary action. Psychoneuroendocrinology. 1984;**9**:97-106

[75] Racagni G, Apud JA, Cocchi D, Locatelli V, Müller EE. GABAergic control of anterior pituitary secretion. Life Sciences. 1982;**31**:823-838

[76] Gamse R, Vaccaro DE, Gamse G, DiPace M, Fox TO, Leeman SE. Release of immunoreactive somatostatin from hypothalamic cells in culture: Inhibition by gamma-aminobutyric acid. Proceedings of the National Academy of Sciences of the United States of America. 1980;**77**:5552-5556

[77] Rage F, Benyassi A, Arancibia S, Tapia-Arancibia L. Gamma-aminobutyric acid-glutamate interaction in the control of somatostatin release from hypothalamic neurons in primary culture: In vivo corroboration. Endocrinology. 1992;**130**:1056-1062

[78] Friedman SD, Baker LD, Borson S, Jensen JE, Barsness S, Craft S, et al. Growth hormone-releasing hormone effects on brain gamma-aminobutyric acid levels in mild cognitive impairment and healthy aging. JAMA Neurology. 2013;**70**:883-890

[79] Tang S, Luo Z, Qiu X, Zhang Y, Lu X, Huang H, et al. Interactions between GHRH and GABAARs in the brains of patients with epilepsy and in animal models of epilepsy. Scientific Reports. 2017;**7**:18110

[80] Cuttica CM, Giusti M, Bocca L, Sessarego P, De Martini D, Valenti S, et al. Nitric oxide modulates in vivo and in vitro growth hormone release in acromegaly. Neuroendocrinology. 1997;**66**:426-431

[81] Fisker S, Nielsen S, Ebdrup L, Bech JN, Christiansen JS, Pedersen EB, et al. The role of nitric oxide in L-arginine-stimulated growth hormone release. Journal of Endocrinological Investigation. 1999;**22**:89-93

[82] Valverde I, Peñalva I, Ghigo E, Casanueva FF, Diéguez C. Involvement of nitric oxide in the regulation of growth hormone secretion in dogs. Neuroendocrinology. 2001;**74**:213-219

[83] Aguila MC. Growth hormone-releasing factor increases somatostatin release and mRNA levels in the rat periventricular nucleus via nitric oxide oxidation. Proceedings of the National Academy of Sciences of the United States of America. 1994;**91**:782

[84] Rubinek T, Rubinfeld H, Hadani M, Barkai G, Shimon I. Nitric oxide stimulates growth hormone secretion from human fetal pituitaries and cultured pituitary adenomas. Endocrine. 2005;**28**:209-216

[85] Rodríguez-Pacheco F, Luque RM, Tena-Sempere M, Malagón MM, Castaño JP. Ghrelin induces growth hormone secretion

via a nitric oxide/cGMP signalling pathway. Journal of Neuroendocrinology. 2008;**20**:406-412

[86] Akril H, Watson SJ, Young E, Lewis ME, Khachaturian H, Walker JM. Endogenous opioids: Biology and function. Annual Review of Neuroscience. 1984;7:223-255

[87] Zadina JE, Hackler L, Ge LJ, Kastin AJ. A potent and selective agonist for the μ-opiate receptor. Nature. 1997;**386**:499-502

[88] Bruhn TO, Tresco PA, Mueller GP, Jackson IM. ß-Endorphin mediates clonidine stimulated growth hormone release. Neuroendocrinology. 1989;**50**:460-463

[89] Wehrenberger WB, Bloch B, Ling N. Pituitary secretion of growth hormone in response to opioid peptides and opiates is mediated through growth hormone-releasing factor. Neuroendocrinology. 1985;**41**:13-16

[90] Borer KT, Nicoski DR, Owens V. Alteration of pulsatile growth hormone secretion by growth-inducing exercise: Involvement of endogenous opiates and somatostatin. Endocrinology. 1986;**118**:844-850

[91] Wuong C, Van Uum SHM, O'Dell L, Lutfy K, Friedman TC. The effects of opioids and opioids analogs on animal and human endocrine systems. Endocrine Reviews. 2010;**31**:98-132

[92] Owen OE, Morgan AP, Kemp HG, Sullivan JM, Herrera MG, Cahill GF Jr. Brain metabolism during fasting. The Journal of Clinical Investigation. 1967;**46**:1589-1595

[93] Stanley S, Domingos AI, Kelly L, Garfield A, Damanpour S, Heisler L, et al. Profiling of glucose-sensing neurons reveals that GHRH neurons are activated by hypoglycemia. Cell Metabolism. 2013;**18**:596-607

[94] Steyn FJ, Huang L, Ngo ST, Leong JW, Tan HY, Xie TY, et al. Development of a method for the determination of pulsatile growth hormone secretion in mice. Endocrinology. 2011;**152**:3165-3171

[95] Steyn FJ, Leong JW, Huang L, Tan HY, Xie TY, Nelson C, et al. GH does not modulate the early fasting-induced release of free fatty acids in mice. Endocrinology. 2012;**153**:273-282

[96] Huang L, Tan HY, Fogarty MJ, Andrews ZB, Veldhuis JD, Herzog H, et al. Actions of NPY, and its Y1 and Y2 receptors on pulsatile growth hormone secretion during the fed and fasted state. The Journal of Neuroscience. 2014;**34**:16309-16319

[97] Patel YC, Wheatley T, Zingg HH. Increased blood somatostatin concentration in streptozotocin diabetic rats. Life Sciences. 1980;**27**:1563-1570

[98] Olchovsky D, Bruno JF, Wood TL, Gelato MC, Leidy JW Jr, Gilbert JM Jr, et al. Altered pituitary growth hormone [GH] regulation in streptozotocin-diabetic rats: A combined defect of hypothalamic somatostatin and GH-releasing factor. Endocrinology. 1990;**126**:53-61

[99] Asplin CM, Faria AC, Carlsen EC, Vaccaro VA, Barr RE, Iranmanesh A, et al. Alterations in the pulsatile mode of growth hormone release in men and women with insulin-dependent diabetes mellitus. The Journal of Clinical Endocrinology and Metabolism. 1989;**69**:239-245

[100] Hayford JT, Danney MM, Hendrix JA, Thompson RG. Integrated concentration of growth hormone in juvenile-onset diabetes. Diabetes. 1980;**29**:391-398

[101] Pietschmann P, Schernthaner G, Prskavec F, Gisinger C, Freyler H. No evidence for increased growth hormone responses to growth hormone-releasing hormone in patients with diabetic retinopathy. Diabetes. 1987;**36**:159-162

[102] Giustina A, Bresciani E, Tassi C, Girelli A, Valentini U. Effect of pyridostigmine on the growth hormone response to growth hormone-releasing hormone in lean and obese type II diabetic patients. Metabolism. 1994;**43**:893-898

[103] Kopelman PG, Mason AC, Noonan K, Monson JP. Growth hormone response to growth hormone releasing factor in diabetic men. Clinical Endocrinology. 1988;**28**:33-38

[104] Richards NT, Wood SM, Christofides ND, Bhuttacharji SC, Bloom SR. Impaired growth hormone response to human pancreatic growth hormone releasing factor [GRF[1-44]] in type 2 [non-insulin-dependent] diabetes. Diabetologia. 1984;**27**:529-553

[105] Imaki T, Shibasaki T, Shizume K, Masuda A, Hotta M, Kiyosawa Y, et al. The effect of free fatty acids on growth hormone-releasing hormone-mediated GH secretion in man. The Journal of Clinical Endocrinology and Metabolism. 1985;**60**:290-293

[106] Casanueva FF, Villanueva L, Dieguez C, Díaz Y, Cabranes JA, Szoke B, et al. Free fatty acids [FFA] block GHRH-stimulated GH secretion in man directly at the pituitary. The Journal of Clinical Endocrinology and Metabolism. 1987;**65**:634-640

[107] Peñalva A, Gaztambide S, Vazquez JA, Dieguez C, Casanueva FF. Role of cholinergic muscarinic pathways on the free fatty acid inhibition of GH responses to GHRH in normal men. Clinical Endocrinology. 1990;**33**:171-176

[108] Hartman ML, Veldhuis JD, Johnson ML, Lee MM, Alberti KG, Samojlik E, et al. Augmented growth hormone [GH] secretory burst frequency and amplitude mediate enhanced GH secretion during a two-day fast in normal men. The Journal of Clinical Endocrinology and Metabolism. 1992;**74**:757-765

[109] Ho KY, Veldhuis JD, Johnson ML, Furlanetto R, Evans WS, Alberti K, et al. Fasting enhances growth-hormone secretion and amplifies the complex rhythms of growth-hormone secretion in man. The Journal of Clinical Investigation. 1988;**81**:968-975

[110] Nam SY, Lee EJ, Kim KR, Lee HC, Nam MS, Cho JH, et al. Long-term administration of acipimox potentiates growth hormone response to growth hormone-releasing hormone by decreasing serum free fatty acid in obesity. Metabolism. 1996;**45**:594-597

[111] Pontiroli AE, Manzoni MF, Malighetti ME, Lanzi R. Restoration of growth hormone [GH] response to GH-releasing hormone in elderly and obese subjects by acute pharmacological reduction of plasma free fatty acids. The Journal of Clinical Endocrinology and Metabolism. 1996;**81**:3998-4001

[112] Alba-Roth J, Albrecht Muller O, Schopohl J, Von Werder K. Arginine stimulates growth hormone secretion by suppressing endogenous somatostatin secretion. The Journal of Clinical Endocrinology and Metabolism. 1988;**67**:1186-1189

[113] Page MD, Diéguez C, Valcavi R, Edwards C, Hall R, Scanlon M. Growth hormone, GH response to arginine and L-Dopa alone and after GHRH treatment. Clinical Endocrinology. 1988;**25**:551

[114] Kelijman M, Frohman LA. The role of the cholinergic pathway in growth hormone feedback. The Journal of

Clinical Endocrinology and Metabolism. 1991;**72**:1081-1087

[115] Frohman LA, Jansson JO. Growth hormone-releasing hormone. Endocrine Reviews. 1986;**7**:223-257

[116] Katakami H, Downs TR, Frohman LA. Decreased hypothalamic growth hormonereleasing hormone content and pituitary responsiveness in hypothyroidism. The Journal of Clinical Investigation. 1986;**77**:1704-1711

[117] Berelowitz M, Maeda K, Harris S, Frohman LA. The effects of alterations of the pituitary-thyroid axis on the hypothalamic content and in vitro release of somatostatin-like immunoreactivity. Endocrinology. 1980;**107**:24

[118] Jones PM, Burrin JM, Ghatei MA, O'Halloran DJ, Legon S, Bloom SR. The influence of thyroid hormone status on the hypothalamo-hypophyseal growth hormone axis. Endocrinology. 1990;**126**:1374

[119] Peterfreund RA, Sawchenko P, Vale W. Thyroid hormones reversely suppress somatostatin secretion and immunoreactivity in cultured neocortical cells. Brain Research. 1985;**328**:259

[120] Devesa J, Lima L, Arce V, Díaz MJ, Tresguerres JAF. State of the art in GH control. In: Pombo M, Rosenfeld RG, editors. Two Decades of Experience in Growth. New York: Raven Press; 1993

[121] Baron J, Huang Z, Oerter KE, Bacher JD, Cuttle GB Jr. Dexamethasone acts locally to inhibit longitudinal bone growth in rabbits. The American Journal of Physiology. 1992;**263**:E489

[122] Baron J, Oerter KE, Colli MJ, Yanovski JA, Novosad JA, Bacher JD, et al. Catch-up of growth after glucocorticoid excess: A mechanism intrinsic to growth plate. Endocrinology. 1994;**135**:1367

[123] Burguera B, Muruais C, Peñalva A, Dieguez C, Casanueva FF. Dual and selective action of glucocorticoid upon basal and stimulated growth hormone release in man. Neuroendocrinology. 1990;**51**:51

[124] Casanueva FF, Burguera B, Tomé MA, Lima L, Tresguerres JA, Devesa J, et al. Depending on dose and time of the administration, dexamethasone potentiates or blocks growth hormone-releasing hormone-induced growth hormone release in man. Neuroendocrinology. 1988;**47**:46

[125] Isaacs RE, Findel RE, Mellon P, Wilson CB, Baxter JD. Hormonal regulation of expression of endogenous and transfected human growth hormone gene. Molecular Endocrinology. 1987;**1**:569

[126] Pack I, Axel R. Glucocorticoids enhance stability of human growth hormone mRNA. Molecular and Cellular Biology. 1987;**7**:1496

[127] Tannenbaum GS, Martin JB. Evidence for an endogenous ultradian rhythm governing growth hormone secretion in the rat. Endocrinology. 1976;**98**:562-570

[128] Jansson JO, Eden S, Isaksson O. Sexual dimorphism in the control of growth hormone secretion. Endocrine Reviews. 1985;**6**:128-150

[129] Steyn FJ, Tolle V, Chen C, Epelbaum J. Neuroendocrine regulation of growth hormone secretion. Comprehensive Physiology. 2016;**6**:687-735

[130] Martha PM Jr, Rogol AD, Veldhuis JD, Kerrigan JR, Goodman DW, Blizzard RM. Alterations in the pulsatile

properties of circulating growth hormone concentrations during puberty in boys. The Journal of Clinical Endocrinology and Metabolism. 1989;**69**:563-570

[131] Rose SR, Municchi G, Barnes KM, Kamp GA, Uriarte MM, Ross JL, et al. Spontaneous growth hormone secretion increases during puberty in normal girls and boys. The Journal of Clinical Endocrinology and Metabolism. 1991;**73**:428-435

[132] Werner H, Koch Y, Baldino F, Gozes I. Steroid regulation of somatostatin mRNA in the rat hypothalamus. The Journal of Biological Chemistry. 1988;**263**:7666-7671

[133] McEwen R. Binding and metabolism of sex steroids by the hypothalamicpituitary unit: Physiological implications. Annual Review of Physiology. 1980;**42**:97-110

[134] Collins S, Caron MG, Lefkowitz RJ. Regulation of adrenergic receptor responsiveness through modulation of receptor gene expression. Annual Review of Physiology. 1991;**53**:497-508

[135] Kerrigan JR, Rogol AD. The impact of gonadal steroid hormone action on growth hormone secretion during childhood and adolescence. Endocrine Reviews. 1992;**13**:281-298

[136] Veldhuis JD, Roemmich JN, Rogol AD. Gender and sexual maturation-dependent contrasts in the neuroregulation of growth hormone secretion in prepubertal and late adolescent males and females—A general clinical research center-based study. The Journal of Clinical Endocrinology and Metabolism. 2000;**85**:2385-2394

[137] García-Galiano D, Cara AL, Tata Z, Allen SJ, Myers MG Jr, Scipani E, et al. ER_α Signaling in GHRH/Kiss1 dual-phenotype neurons plays sex-specific roles in growth and puberty. The Journal of Neuroscience. 2020;**40**:9455-9466

[138] Faria AC, Bekenstein LW, Booth RA Jr, Vaccaro VA, Asplin CM, Veldhuis JD, et al. Pulsatile growth hormone release in normal women during the menstrual cycle. Clinical Endocrinology. 1992;**36**:591-596

[139] Wilson EE, Word RA, Byrd W, Carr BR. Effect of superovulation with human menopausal gonadotropins on growth hormone levels in women. The Journal of Clinical Endocrinology and Metabolism. 1991;**73**:511-515

[140] Frantz AG, Rabkin MT. Effects of estrogen and sex difference on secretion of human growth hormone. The Journal of Clinical Endocrinology and Metabolism. 1965;**25**:1470-1480

[141] Jansson JO, Frohman LA. Inhibitory effect of the ovaries on neonatal androgen imprinting of growth hormone secretion in female rats. Endocrinology. 1987;**121**:1417-1423

[142] Mauras N, Blizzard RM, Link K, Johnson ML, Rogol AD, Veldhuis JD. Augmentation of growth hormone secretion during puberty: Evidence for a pulse-amplitude modulated phenomenon. The Journal of Clinical Endocrinology and Metabolism. 1987;**64**:596-601

[143] Wiedemann E, Schwartz E, Frantz AG. Acute and chronic estrogen effects upon serum somatomedin activity, growth hormone, and prolactin in man. The Journal of Clinical Endocrinology and Metabolism. 1976;**42**:942-952

[144] Bellantoni MF, Harman SM, Cho DE, Blackman MR. Effects of progestin-opposed transdermal estrogen administration on growth hormone and insulin-like growth factor-I in postmenopausal women of different ages. The Journal of Clinical Endocrinology and Metabolism. 1991;**72**:172-178

[145] Bellantoni MF, Vittone J, Campfield AT, Bass KM, Harman SM,

Blackman MR. Effects of oral versus transdermal estrogen on the growth hormone/insulin-like growth factor I axis in younger and older post-menopausal women: A clinical research center study. The Journal of Clinical Endocrinology and Metabolism. 1996;**81**:2848-2853

[146] Weissberger AJ, Ho KK, Lazarus L. Contrasting effects of oral and transdermal routes of estrogen replacement therapy on 24-hour growth hormone [GH] secretion, insulin-like growth factor I, and GH-binding protein in postmenopausal women. The Journal of Clinical Endocrinology and Metabolism. 1991;**72**:374-381

[147] Veldhuis JD, Erickson D, Yang R, Takahashi P, Bowers C. Endogenous estrogen regulates somatostatin-induced rebound GH secretion in postmenopausal women. The Journal of Clinical Endocrinology and Metabolism. 2016;**101**:4298-4304

[148] Lima L, Arce V, Lois N, Fraga C, Lechuga MJ, Tresguerres JAF, et al. Growth hormone [GH] responsiveness to GHRH in normal adults is not affected by short-term gonadal blockade. Acta Endocrinologica. 1989;**120**:31-36

[149] Vijayakumar A, Novosyadlyy R, Wu Y, Yakar S, LeRoith D. Biological effects of growth hormone on carbohydrate and lipid metabolism. Growth Hormone & IGF Research. 2010;**2010**(20):1-7

[150] Kelley DE, Wing R, Buonocore C, Sturis J, Polonsky K, Fitzsimmons M. Relative effects of calorie restriction and weight loss in noninsulin-dependent diabetes mellitus. The Journal of Clinical Endocrinology and Metabolism. 1993;**77**:1287-1293

[151] De Marinis L, Bianchi A, Mancini A, Gentilella R, Perrelli M, Giampietro A, et al. Growth hormone secretion and leptin in morbid obesity before and after biliopancreatic diversion: Relationships with insulin and body composition. The Journal of Clinical Endocrinology and Metabolism. 2004;**89**:174-180

[152] Wagner C, Caplan SR, Tannenbaum GS. Genesis of the ultradian rhythm of GH secretion: A new model unifying experimental observations in rats. The American Journal of Physiology. 1998;**275**:E1046-E1054

[153] Wakabayashi I, Demura R, Miki N, Ohmura E, Miyoshi H, Shizume K. Failure of naloxone to influence plasma growth hormone, prolactin, and cortisol secretions induced by insulin hypoglycemia. The Journal of Clinical Endocrinology and Metabolism. 1980;**1980**(50):597-599

[154] Luque RM, Kineman RD. Impact of obesity on the growth hormone axis: Evidence for a direct inhibitory effect of hyperinsulinemia on pituitary function. Endocrinology. 2006;**147**:2754-2763

[155] Melmed S. Insulin suppresses growth hormone secretion by rat pituitary cells. The Journal of Clinical Investigation. 1984;**73**:1425-1433

[156] Gahete MD, Cordoba-Chacon J, Lin Q, Bruning JC, Kahn CR, Castano JP. Insulin and IGF-I inhibit GH synthesis and release in vitro and in vivo by separate mechanisms. Endocrinology. 2013;**154**:2410-2420

[157] Folli F, Ghidella S, Bonfanti L, Kahn CR, Merighi A. The early intracellular signaling pathway for the insulin/insulin-like growth factor receptor family in the mammalian central nervous system. Molecular Neurobiology. 1996;**13**:155-183

[158] Hill JM, Lesniak MA, Pert CB, Roth J. Autoradiographic localization of insulin receptors in rat brain: Prominence in olfactory and limbic areas. Neuroscience. 1986;**17**:1127-1138

[159] Kar S, Chabot JG, Quirion R. Quantitative autoradiographic localization of [125I] insulin-like growth factor I, [125I] insulin-like growth factor II, and [125I] insulin receptor binding sites in developing and adult rat brain. The Journal of Comparative Neurology. 1993;**333**:375-397

[160] Werther GA, Hogg A, Oldfield BJ, McKinley MJ, Figdor R, Allen AM, et al. Localization and characterization of insulin receptors in rat brain and pituitary gland using in vitro autoradiography and computerized densitometry. Endocrinology. 1987;**121**:1562-1570

[161] Becker K, Stegenga S, Conway S. Role of insulin-like growth factor I in regulating growth hormone-release and feedback in the male rat. Neuroendocrinology. 1995;**61**:573-583

[162] Bohannon NJ, Figlewicz DP, Corp ES, Wilcox BJ, Porte D Jr, Baskin DG. Identification of binding sites for an insulin-like growth factor, IGF-I in the median eminence of the rat brain by quantitative autoradiography. Endocrinology. 1986;**119**:943-945

[163] Soto JL, Castrillo JL, Domínguez F, Diéguez C. Regulation of the pituitary-specific transcription factor GHF-1/Pit-1 messenger ribonucleic acid levels by growth hormone secretagogues in rat anterior pituitary cells in monolayer culture. Endocrinology. 1995;**136**:3863-3870

[164] Bowers CY. History to the discovery of ghrelin. Methods in Enzymology. 2012;**514**:3-32

[165] Bowers CY, Reynolds GA, Durham D, Barrera CM, Pezzoli SS, Thorner MO. Growth hormone [GH]-releasing peptide stimulates GH release in normal men and acts synergistically with GH-releasing hormone. The Journal of Clinical Endocrinology and Metabolism. 1990;**70**:975-982

[166] Howard AD, Feighner SD, Cully DF, Arena JP, Liberator PA, Rosenblum CI. A receptor in pituitary and hypothalamus that functions in growth hormone release. Science. 1996;**273**:974-977

[167] Kojima M, Hosoda H, Date Y, Nakazato M, Matsuo H, Kangawa K. Ghrelin is a growth-hormone-releasing acylated peptide from stomach. Nature. 1999;**402**:656-660

[168] Devesa J. The complex world of regulation of pituitary growth hormone secretion: The role of ghrelin, klotho, and Nesfatins in it. Frontiers in Endocrinology. 2021;**12**:636403

[169] Kojima M, Kangawa K. Ghrelin: Structure and function. Physiological Reviews. 2005;**85**:495-522

[170] Rigamonti AE, Pincelli AI, Corrá B, Viarengo R, Bonomo SM, Galimberti D, et al. Plasma ghrelin concentrations in elderly subjects: Comparison with anorexic and obese patients. The Journal of Endocrinology. 2002;**175**:R1-R5

[171] Caicedo D, Díaz O, Devesa P, Devesa J. Growth hormone [GH] and cardiovascular system. International Journal of Molecular Sciences. 2018;**19**:290

[172] Baldanzi G, Filigheddu N, Cutrupi S, Catapano F, Bonisoni S, Fubini A, et al. Ghrelin and des-acyl ghrelin inhibit cell death in cardiomyocytes and endothelial cells through ERK1/2 and PI 3-kinase/AKT. The Journal of Cell Biology. 2002;**159**:1029-1037

[173] Gutierrez JA, Solenberg PJ, Perkins DR, Willency JA, Knierman MD, Jin Z, et al. Ghrelin octanoylation mediated by an orphan lipid transferase. Proceedings of the National Academy of Sciences of the United States of America. 2008;**195**:6320-6325

[174] Yang J, Brown MS, Liang G, Grishin NV, Goldstein JL. Identification of the acyltransferase that octanoylates ghrelin, an appetite-stimulating peptide hormone. Cell. 2008;**132**:387-396

[175] Dimaraki EV, Jaffe CA, Demott-Friberg R, Russell-Aulet M, Bowers CY, Marbach P, et al. Generation of growth hormone pulsatility in women: Evidence against somatostatin withdrawal as pulse initiator. American Journal of Physiology. Endocrinology and Metabolism. 2001;**280**:E489-E495

[176] Dimaraki EV, Jaffe CA, Bowers CY, Marbach P, Barkan AL. Pulsatile and nocturnal growth hormone secretions in men do not require periodic declines of somatostatin. American Journal of Physiology. Endocrinology and Metabolism. 2003;**285**:E163-E170

[177] Maheshwari HG, Pezzoli SS, Rahim A, Shalet SM, Thorner MO, Baumann G. Pulsatile growth hormone secretion persists in genetic growth hormone-releasing hormone resistance. American Journal of Physiology. Endocrinology and Metabolism. 2002:E943-E951

[178] Pantel J, Legendre M, Cabrol S, Hilal L, Hajaji Y, Morisset S, et al. Loss of constitutive activity of the growth hormone secretagogue receptor in familial short stature. The Journal of Clinical Investigation. 2006;**116**:760-768

[179] Tannenbaum GS, Epelbaum J, Bowers CY. Interrelationship between the novel peptide ghrelin and somatostatin/growth hormone-releasing hormone in regulation of pulsatile growth hormone secretion. Endocrinology. 2003;**144**:967-974

[180] Root A, Root M. Clinical pharmacology of human growth hormone and its secretagogues. Current Drug Targets. Immune, Endocrine and Metabolic Disorders. 2002;**2**:27-52

[181] Korbonits M, Kojima M, Kangawa K, Grossman AB. Presence of ghrelin in normal and adenomatous human pituitary. Endocrine. 2001;**14**:101-104

[182] Korbonits M, Bustin SA, Kojima M, Jordan S, Adams EF, Lowe DG, et al. The expression of the growth hormone secretagogue receptor ligand ghrelin in normal and abnormal human pituitary and other neuroendocrine tumors. The Journal of Clinical Endocrinology and Metabolism. 2001;**86**:881-887

[183] Kamegai J, Tamura H, Shimizu T, Ishii S, Tatsuguchi A, Sugihara H, et al. The role of pituitary ghrelin in growth hormone [GH] secretion: GH-releasing hormone-dependent regulation of pituitary ghrelin gene expression and peptide content. Endocrinology. 2004;**145**:3731-3738

[184] Magdaleno-Méndez A, Domínguez B, Rodríguez-Andrade A, Barrientos-Morales M, Cervantes Acosta P, Hernández-Beltrán A, et al. Ghrelin increases growth hormone production and functional expression of NaV1.1 and NaV1.2 channels in pituitary somatotropes. Endocrine. 2015;**48**:929-936

[185] Vázquez MJ, Novelle MG, Rodríguez-Pacheco F, Lage R, Varela L, López M, et al. AMPK-dependent mechanisms but not hypothalamic lipid signaling mediates GH-secretory responses to GHRH and ghrelin. Cell. 2020;**9**:1940

[186] Devesa J, Lima L, Tresguerres JA. Neuroendocrine control of growth hormone secretion in humans. Trends in Endocrinology and Metabolism. 1992;**3**:175-183

[187] Nass R, Liu J, Pezzoli S, Oliveri M, Gaylinn B, Thorner M. 24-h mean acyl-ghrelin levels are decreased in older adults. In: 4th International Congress of the GRS and the IGF Society. Italy: Springer Link; 2008. p. 51

[188] Broglio F, Benso A, Castiglioni C, Gottero C, Prodam F, Destefanis S, et al. The endocrine response to ghrelin as a function of gender in humans in young and elderly subjects. The Journal of Clinical Endocrinology and Metabolism. 2003;**88**:1537-1542

[189] Cummings DE, Purnell JQ, Frayo RS, Schmidova K, Wisse BE, Weigle DS. A preprandial rise in plasma ghrelin suggests a role in meal initiation in humans. Diabetes. 2001;**50**:1714-1719

[190] Lee H-M, Wang G, Englander EW, Kojima M, Greeley GH Jr. Ghrelin, a new gastrointestinal endocrine peptide that stimulates insulin secretion: Enteric distribution, ontogeny, influence of endocrine, and dietary manipulations. Endocrinology. 2002;**143**:185-190

[191] Zhao T-J, Sakata I, Lin Li R, Liang G, Richardson JA, Brown MS, et al. Ghrelin secretion stimulated by ß1-adrenergic receptors in cultured ghrelinoma cells and in fasted mice. Proceedings of the National Academy of Sciences of the United States of America. 2010;**107**:15868-15873

[192] Sugino T, Yamaura J, Yamagishi M, Kurose Y, Kojima M, Kangawa K, et al. Involvement of cholinergic neurons in the regulation of the ghrelin secretory response to feeding in sheep. Biochemical and Biophysical Research Communications. 2003;**304**:308-312

[193] Yin X, Li Y, Xu G, An W, Zhang W. Ghrelin fluctuation, what determines its production? Acta Biochimica et Biophysica Sinica. 2009;**41**:188-197

[194] de la Cruz DC, Norlén P, Hakanson R. Secretion of ghelin from rat stomach ghrelin cells in response to local microinfusion of candidate messenger compounds: A microdialysis study. Regulatory Peptides. 2007;**143**:118-126

[195] Nonogaki K. Ghrelin and feedback systems. Vitamins and Hormones. 2008;**77**:149-170

[196] Arosio M, Ronchi CL, Gebbia C, Cappiello V, Beck-Peccoz P, Peracchi M. Stimulatory effects of ghrelin on circulating somatostatin and pancreatic polypeptide levels. The Journal of Clinical Endocrinology and Metabolism. 2003;**88**:701-704

[197] Chabot F, Caron A, Laplante M, St-Pierre DH. Interrelationships between ghrelin, insulin and glucose homeostasis: Physiological relevance. World Journal of Diabetes. 2014;**5**:328-341

[198] Broglio F, Arvat E, Benso A, Gottero C, Muccioli G, Papotti M, et al. Ghrelin, a natural GH secretagogue produced by the stomach, induces hyperglycemia and reduces insulin secretion in humans. The Journal of Clinical Endocrinology and Metabolism. 2001;**86**:5083-5086

[199] Kalra SP, Ueno N, Kalra PS. Stimulation of appetite by ghrelin is regulated by leptin restraint: Peripheral and central sites of action. The Journal of Nutrition. 2005;**135**:1331-1335

[200] Pöykkö SM, Ukkola O, Kauma H, Kellokoski E, Hörkkö S, Kesäniemi YA. The negative association between plasma ghrelin and IGF-I is modified by obesity, insulin resistance and type 2 diabetes. Diabetologia. 2005;**48**:309-316

[201] Whatmore AJ, Hall CM, Jones J, Westwood M, Clayton PE. Ghrelin concentrations in healthy children and adolescents. Clinical Endocrinology. 2003;**59**:649-654

[202] Bellone S, Rapa A, Vivenza D, Castellino N, Petri A, Bellone J, et al. Circulating ghrelin levels as function of gender, pubertal status and adiposity in childhood. Journal of Endocrinological Investigation. 2002;**25**:RC13-RC15

[203] Stawerska R, Smyczyriska J, Hilczer M, Lewinski A. Relationship between IGF-I concentration and metabolic profile in children with growth hormone deficiency: The influence of children's nutritional state as the ghrelin, leptin, adiponectin, and resistin serum concentrations. International Journal of Endocrinology. 2017;**2017**:5713249

[204] Kuro-O M, Matsumura Y, Aizawa H, Kawaguchi H, Suga T, Utsugi T, et al. Mutation of the mouse klotho gene leads to a syndrome resembling ageing. Nature. 1997;**390**:45-51

[205] Chen CD, Podvin S, Gillespie E, Leeman SE, Abraham CR. Insulin stimulates the cleavage of and release of the extracellular domain of Klotho by ADAM10 and ADAM 17. Proceedings of the National Academy of Sciences of the United States of America. 2007;**104**:19796-19801

[206] Rubinek T, Shahmoon S, Shabtay-Orbach A, Ben Ami M, Levy-Shraga Y, Mazor-Aronovitch K, et al. Klotho response to treatment with growth hormone and the role of IGF-I as a mediator. Metabolism. 2016;**65**:1597-1604

[207] Kurosu H, Yamamoto M, Clark JD, Pastor JV, Nandi A, Gurnani P, et al. Suppression of aging in mice by the hormone Klotho. Science. 2005;**309**:1829-1833

[208] Shahmoon S, Rubinfeld H, Wolf I, Cohen ZR, Hadami M, Shimon H, et al. The aging suppressor klotho: A potential regulator of growth hormone secretion. American Journal of Physiology. Endocrinology and Metabolism. 2014;**307**:E3326-EE334

[209] Danger JM, Tonon MC, Jenks BG, Saint-Pierre S, Martel JC, Fasolo A, et al. Neuropeptide Y: Localization in the central nervous system and neuroendocrine functions. Fundamental & Clinical Pharmacology. 1990;**4**:307-340

[210] Pierroz DD, Catzeflis C, Aebi AC, Rivier JE, Aubert ML. Chronic administration of neuropeptide Y into the lateral ventricle inhibits both the pituitary-testicular axis and growth hormone and insulin-like growth factor I secretion in intact adult male rats. Endocrinology. 1996;**137**:3-12

[211] Hisano S, Tsuruo Y, Kagotani Y, Daikoku S, Chihara K. Immuno histochemical evidence for synaptic connections between neuropeptide Y-containing axons and periventricular somatostatin neurons in the anterior hypothalamus in rats. Brain Research. 1990;**520**:170-177

[212] Balasubramaniam A, Joshi R, Su C, Friend LA, James JH. Neuropeptide Y (NPY) Y2 receptor-selective agonist inhibits food intake and promotes fat metabolism in mice: Combined anorectic effects of Y2 and Y4 receptor-selective agonists. Peptides. 2007;**28**:235-240

[213] Keen-Rhinehart E, Bartness TJ. NPY Y1 receptor is involved in ghrelin- and fasting-induced increases in foraging, food hoarding, and food intake. American Journal of Physiology. Regulatory, Integrative and Comparative Physiology. 2007;**292**:R1728-R1737

Section 3

Therapeutic Approaches

Chapter 4

Growth Hormone and Insulin-like Growth Factor-1 in Children with Cholestatic Diseases and Pediatric Liver Transplantation

Sergey V. Gautier, Olga P. Shevchenko, Olga M. Tsirulnikova, Rivada M. Kurabekova and Irina E. Pashkova

Abstract

The growth hormone/insulin-like growth factor 1 (GH/IGF-1) system is the most important regulator of growth, regeneration, and metabolism in children and adults. Children with congenital cholestatic diseases have elevated GH blood levels, which is combined with growth failure and body mass deficit. Congenital cholestatic diseases lead to end-stage liver disease (ESLD), where GH bioavailability, mediated through IGF-1, is impaired. Blood IGF-1 levels are decreased due to impaired production by the liver. This study included 148 children up to 5 years (60 months) old with congenital cholestatic diseases. The patients underwent liver transplantation (LT) at a leading transplant center in Russia. The clinical significance of the GH/IGF-1 axis in pediatric liver recipients was investigated. Relationship between the patients' GH/IGF-1 levels and anthropometric parameters was analyzed before and after LT. It was shown that LT leads to renewal/recovery of GH-IGF-1 regulation and improved anthropometric parameters (body height and body mass) in pediatric recipients.

Keywords: GH, IGF-1, congenital cholestatic diseases, pediatric liver transplantation, anthropometric parameters

1. Introduction

Liver transplantation (LT) is the only effective treatment for patients with end-stage liver disease (ESLD). The first successful pediatric LT was performed by Thomas Starzl in 1967 on a child with biliary atresia (BA). This marked the beginning of the use of LT in pediatric practice [1, 2]. By 2020, the total number of LTs performed was about 32,000 per year, 7–8% of which were pediatric transplant surgeries [3].

In the Russian Federation, the first pediatric LT was performed in 1997. Today, the total number of transplantations performed has exceeded 1000. The need for this type of medical care is fully satisfied in Russia. The LT program is being implemented at Shumakov National Medical Research Center of Transplantology and Artificial Organs (Shumakov Center), Russia's leading transplant center; over 100 LTs are performed annually. Surgeries are performed in children with extremely low body mass, starting from the first months of life.

Invasive diagnostic procedures, such as needle biopsy, carry high risks for young children with low body mass. Therefore, it is an urgent task to search for markers for the development of noninvasive methods of assessing liver graft function. Given physical developmental delays in children, which are associated with hepatobiliary diseases and subsequent hormonal dysregulation, GH and IGF-1, the key elements in neurohumoral control of liver function can serve as important indicators reflecting the outcome of LT. The significance of GH and IGF-1 in pediatric LT is related to the role of hormones in the regulation of growth and body mass, their influence on hepatocyte function and immune system activity. A number of studies have shown the relationship between the GH/IGF-1 axis and recipient/graft survival [4–6].

Despite the large number of studies on the role of the GH/IGF-1 axis in children and adults, none of them are devoted to their relationship with physical development in children after LT [7–9]. The study of the role of the GH/IGF-1 axis in the neurohumoral regulation of liver graft function in young children is an important task. It may allow to assess the functional ability of the graft by noninvasive methods and will also create a vector for coming up with personalized therapy. This chapter describes the impact of the GH/IGF-1 axis on the outcome of pediatric LT (namely the clinical significance of the GH/IGF-1 axis in pediatric liver recipients) and characterizes its relationship with patients' anthropometric parameters.

2. Growth hormone and insulin-like growth factor 1

GH and IGF-1 are key links in the neurohumoral regulation of overall metabolism and can act through various intracellular signaling pathways. GH stimulates IGF-1 synthesis, which, in turn, affects GH production by the principle of negative feedback, inhibiting its synthesis (**Figure 1**).

The physiological effects of GH and IGF-1 on cells are mediated through transmembrane receptors found on the surface of many cell types, including hepatocytes and lymphocytes [10]. The effect of GH and IGF-1 is largely determined by the level of receptor expression, which depends on cell type and can change under the influence of various factors. The action of the GH/IGF-1 axis depends on GH and IGF-1 production on one hand, and IGF-binding proteins, proteases that degrade the IGF-binding proteins complex, and GH and IGF-1 receptors, on the other hand [11].

GH is synthesized in the anterior pituitary cells and secreted into the blood, with peak blood levels every 3–5 hours. The highest blood content of the hormone occurs during fetal development. With age, the baseline, frequency, and amplitude of hormone secretion peaks decrease. The nature of GH secretion differs in men and women and depends on age. The range of baseline GH levels in children aged 1–3 years is 2–10 ng/m, in adults, 1–5 ng/ml [12, 13].

IGF-1 is a polypeptide hormone produced by many tissues, but more than 90% of circulating IGF-1 is synthesized by hepatocytes [3]. Serum IGF-1 levels, in contrast to GH, practically do not change during a day. The range of reference IGF-1 levels in children aged 1–3 years is 5–300 ng/ml; the maximum IGF-1 level in children is observed during puberty and gradually decreases with the years [14, 15].

Growth regulation is one of the main functions that GH and IGF-1 have in common. In addition, GH in both children and adults plays an important role in the regulation of metabolism. IGF-1, being a major mediator of anabolic and mitogenic effects of

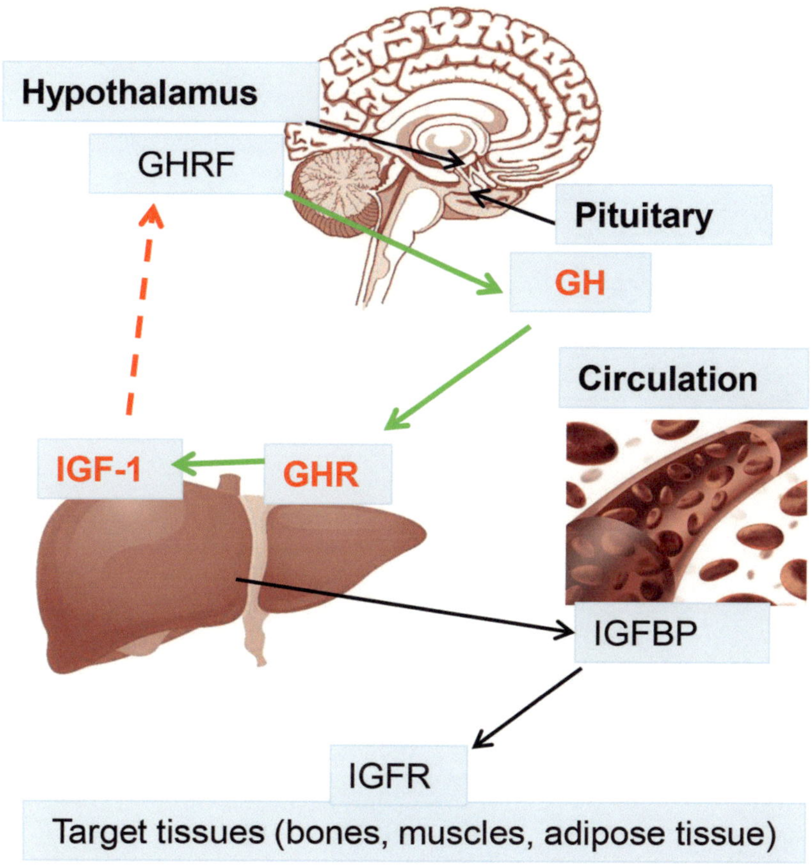

Figure 1.
Scheme of neurohumoral regulation of growth by the GH/IGF-1 system. GHRF – growth hormone releasing factor, GH – growth hormone, GHR - growth hormone receptor, IGF-1 – insulin-like growth factor 1, IGFBP – IGF binding protein, IGFR – IGF receptor.

GH in peripheral tissues, is an important factor in body mass regulation. However, GH and IGF-1 have different effects on glucose and lipid metabolism: GH increases blood glucose levels and promotes lipolysis, whereas IGF-1 has opposite effects [16–18].

3. Effect of GH/IGF-1 on liver function

The liver is the largest internal organ in humans; in a child it makes up 4.4% of body mass, which is 1.5 times more than in an adult—2.8%. The importance and diversity of liver functions in young children should be noted: it is not only bile production, general metabolism, detoxification, but also, importantly for pediatric liver recipients, the liver works as an immune organ, maintaining overall immune homeostasis of the body (**Figure 2**).

Humoral regulation of liver function is provided by the peculiarity of the hepatic portal system and the diversity of the cellular composition, which are targets for so many hormones and cytokines (**Figure 3**).

The liver forms a unique microenvironment due to its anatomy, in particular the circulatory system consisting of the portal vein, hepatic artery, and liver sinusoid. Portal

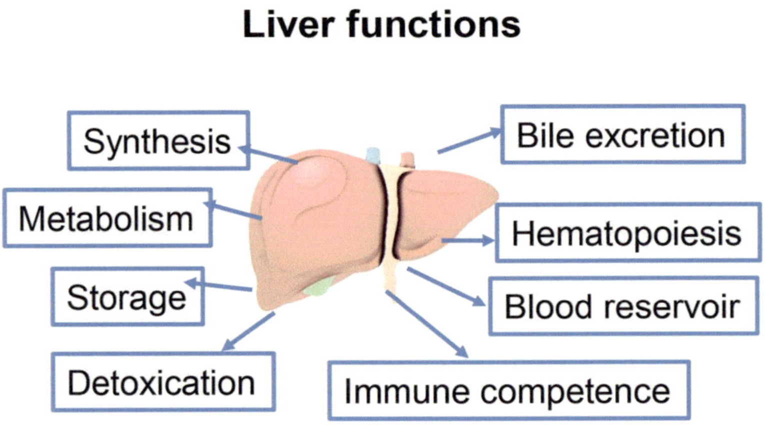

Figure 2.
Liver functions in children.

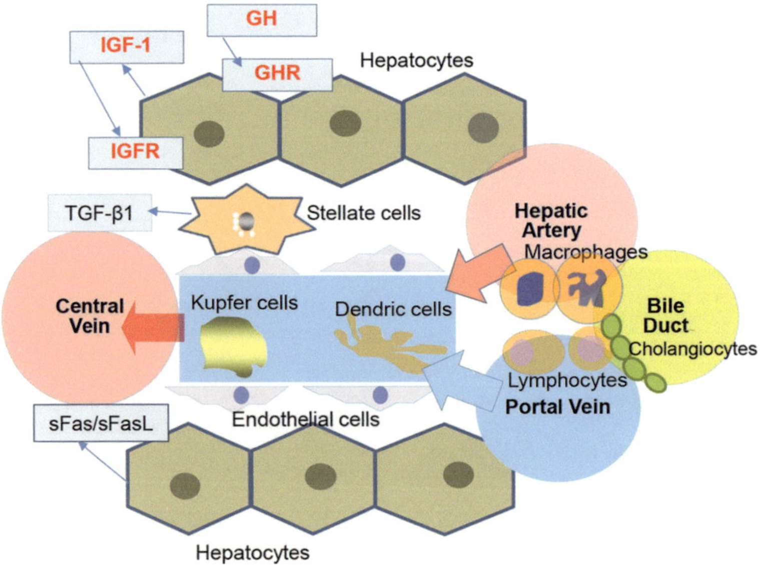

Figure 3.
Peculiarities of liver structure and cells composition. The hepatic blood supply system consists of the portal vein, the hepatic artery, and the sinusoidal capillary system of the liver, which flows into the central vein. Liver cells with antigen-presenting function: dendritic, Kupffer, stellate, and endothelial. GH – growth hormone, GHR – growth hormone receptor, IGF-1 – insulin-like growth factor 1, IGFR – IGF receptor, TGF-beta1 – transforming growth factor beta 1, sFas/sFasL – soluble Fas/soluble Fas ligand.

venous blood is rich in dietary antigens and toxins, to which tolerance is developed by a large set of liver cells with antigen-presenting function: dendritic, Kupffer, stellate, and endothelial cells [19–23]. Liver sinusoidal endothelial cells, which, unlike the vascular endothelium, do not have a basement membrane and tight junctions, are capable of blocking proinflammatory CD8+ T cells [24]. The liver is characterized by the generation of autoantigen-specific regulatory T cells, which, through active immunosuppression of type 1 and 2 T helper cells (Th1/2), form a mechanism of autoimmune tolerance [25–27]. It is also assumed that natural killers play a significant role in the immunocompetent properties of the liver; their number in the human liver is up to 10% of all lymphocytes [28]. Liver stellate cells are also capable of suppressing immune response, causing apoptosis of activated T cells through expression of costimulatory molecules (CD28/B7) B7-H1 and the production of cytokines—interleukins (IL) 6 and 10, transforming growth factor beta (TGF-β) [29]. Stellate cells induce tolerance through apoptosis of activated T cells via the Fas/Fas ligand system, IL-10 and TGF-β [20].

The GH/IGF-1 axis is the most important regulatory system of the body that controls cell growth and is closely related to liver function [4, 5, 30]. Previously, it was believed that IGF-1 does not directly affect the function of hepatocytes, because in a healthy liver, a small number of IGF-1 receptors are expressed on the surface of hepatocytes. However, further studies have shown that their expression is increased in some liver diseases [30]. In acute viral hepatitis, chronic hepatitis C and B, the expression of IGF-1 receptors on hepatocytes is higher than in a healthy liver. Meanwhile, there are elevated IGF-1 levels, which is believed to accelerate the regeneration of damaged hepatocytes [31].

The antifibrotic effect of IGF-1 is realized both directly through the GH/IGF-1 axis and indirectly through regulation of other profibrogenic factors [32, 33]. Stellate cells play a key role in the development of liver fibrosis: their activation caused by chronic trauma, oxidative stress, and increased levels of inflammatory cytokines and lipopolysaccharides leads to transformation into fibroblasts [34]. It has been shown that IGF-1 is able to inactivate hepatic stellate cells and induce their aging, thus limiting the development of fibrosis [35]. GH and IGF-1 regulate lipid metabolism and influence the development of steatosis, inflammation, and liver fibrosis in a certain way [36]. Thus, reduced IGF-1 production in the liver is not only the result of impaired liver function, but also plays an important role in the development of fibrosis.

4. GH/IGF-1 in liver diseases

IGF-1 synthesis by hepatocytes is impaired in liver diseases. This leads to reduced IGF-1 levels and, according to the principle of negative feedback, leads to increased GH secretion. Despite the high serum GH levels in patients with chronic liver diseases, low IGF-1 levels can also be the result of hepatocyte resistance to GH [37, 38]. It is assumed that metabolic disorders—insulin resistance, malnutrition, osteopenia, etc.—often found in patients with liver diseases are associated with impaired neurohumoral regulation and are caused by IGF-1 deficiency [39–41].

The degree of decrease in IGF-1 levels in patients with chronic liver diseases correlates with the severity of hepatocyte dysfunction [36, 42]. Introduction of recombinant IGF-1 stops liver fibrous degeneration [43, 44]. In an animal experiment with nonalcoholic steatohepatitis (NASH), recombinant IGF-1 was shown to improve liver function in cirrhosis. Also in the experiment, it was found that the degree of hepatic ischemia-reperfusion injury is less at higher IGF-1 levels [45, 46]. Experimental studies suggest the possibility of using IGF-1 in clinical practice. Administration of recombinant IGF-1 in patients with liver cirrhosis has been shown to increase serum albumin levels and improve energy metabolism [47].

In adult patients with liver cirrhosis, GH levels are significantly higher than in healthy ones, which is associated with impaired IGF-1 synthesis by the liver in ESLD [17]. Already on day 7 following LT in adults, GH levels decrease to normal values, and those of IGF-1 increases [39].

In recent decades, there has been much focus on the study of the GH/IGF-1 axis as factors associated with longevity and quality of life [48]. It is known that life expectancy is closely related to the GH/IGF-1 axis, which may be of some importance for the development of techniques for predicting recipient and graft survival [6, 49]. It has been shown that IGF-1 can improve survival rates in rats with lipopolysaccharide/D-galactosamine-induced acute liver failure. Prophylactic administration of IGF-1 in the animals prevented an increase in bilirubin levels and transaminase activity [44].

5. GH/IGF-1 in pediatric LT

In pediatric LT, GH and IGF-1 play a significant role in the regulation of growth and body mass, hepatocyte function, and immune system activity [4–6].

Indications for LT in children include ESLD with a life expectancy of less than 1 year, acute liver failure, unresectable liver tumors, and metabolic disease. In 80% of cases, ESLD is caused by congenital anomalies of the biliary tract, such as biliary atresia (BA) and biliary hypoplasia (BH), Alagille syndrome and disease, etc. The incidence of such diseases is about 1 in 10,000–20,000 newborns, atresia is slightly more common in newborn girls than boys. Biliary disorders lead to cholestasis, fibrosis, and cirrhosis, which in turn result in esophageal varices, splenomegaly, hemorrhagic syndrome, esophageal or gastrointestinal bleeding, etc. BA with the inevitable formation of fibrosis and cirrhosis leads to decreased detoxification of metabolites, chronic malnutrition, coagulopathy, growth factor deficiencies, and an excess of proinflammatory cytokines, which together can contribute to delayed physical and psychomotor development. In the absence of early surgical intervention, BA children usually do not live up to one year, dying from liver failure, bleeding, associated pneumonia, cardiovascular disease, and intercurrent infections. With incomplete biliary atresia, some children may live up to 10 years. Therefore, congenital anomalies affecting the biliary tract are a strong indication for LT. LTs performed at Shumakov Center fully cover the needs for this type of care in Russia [50].

The shortage of donor organs for children is compensated by living related donation of liver fragments. As a rule, young children are transplanted with the left lateral lobe of the liver of a living or deceased donor (**Figure 4**).

Transplantation of a liver fragment from a living related donor to children makes it possible to fully satisfy the need for this type of therapy. Improved pre-transplantation preparation, patient selection, organ preservation techniques, better immunosuppressive therapy, and post-transplant follow-up techniques have led to outstanding results in this area.

Unlike other solid organs, the liver is an immune-privileged organ, which is due to the peculiarities of development, blood supply, structure, and function of the liver [22, 51, 52]. After LT, incidence of graft rejection is significantly lower than after transplantation of other solid organs, such as the kidney or heart. The best recipient and graft survival rates are observed in pediatric LT from a living related donor. This is due to the peculiarities of immune response development at an early age, the possibility of ensuring greater tissue compatibility, and a shorter graft ischemia time. Pediatric LT from a living related donor is a model of the relationship between the graft and the recipient's body in conditions of minimal risk of rejection and low immunosuppression.

LT has become a standard therapy for children with ESLD, acute or chronic, as well as liver neoplasms and metabolic disorders. According to world statistics, the 5- and 10-year graft survival in children is 85–90% and 77–80%, respectively [53].

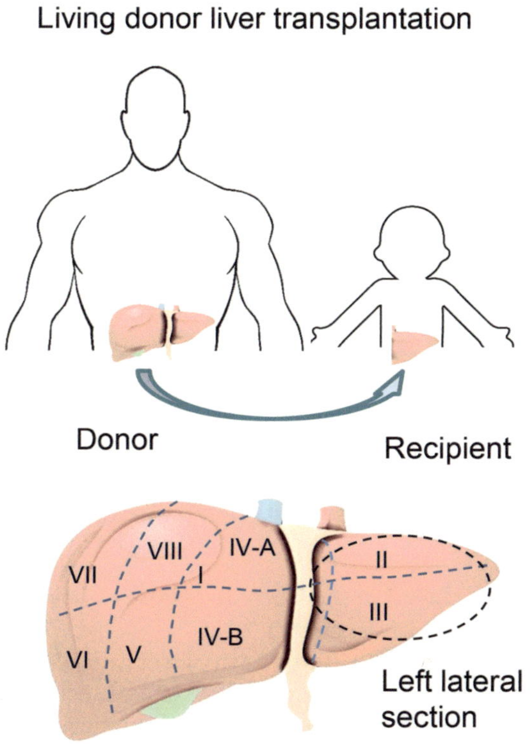

Figure 4.
Transplantation of the left lateral liver section in children from an adult living donor.

The results we achieved at Shumakov Center are fully consistent with global ones. For example, the average 1-year survival rate was 93%, and the 3-year survival rate was up to 90% [50, 54].

5.1 Patients

The present study included 148 children ≤5 years (60 months) old, selected by random sampling, who underwent liver fragment transplantation at Shumakov Center. The study was approved by the Local Ethics Committee of Shumakov Center (Minutes No.28/1d dated April 15, 2016). All the patients or their guardians gave written informed consent to participate in the study.

The mean age of the recipients was 11 ± 9.8 months (median, 8 months). LT was indicated due to ESLD resulting from various pathologies: biliary atresia (BA, n = 86), Byler disease (n = 15), biliary hypoplasia (BH, n = 14), Alagille syndrome (n = 12), Caroli disease and syndrome (n = 5), and others (n = 16). In the "other" group, we included patients with diseases, which did not exceed three cases, such as autoimmune and fulminant hepatitis, Von Gierke disease, portal vein developmental anomaly, Crigler-Najjar syndrome, Budd-Chiari syndrome, hepatoblastoma, alpha-1 antitrypsin deficiency, and previous liver transplant dysfunction (**Table 1**).

The comparison group consisted of 16 children, 9 boys and 7 girls, practically healthy, examined after treatment of intestinal dysbacteriosis, without other diseases, aged 3–25 (median, 12) months, anthropometric parameters (body height, body mass) were in the range of average population values (25th–75th percentile). The age

Characteristics	Value	Healthy children	*p* value
Number of patients, n	148	16	-
Age, median, (range), month	8 (2-60)	12 (3-25)	0,78
Gender (M/F) (n; %)	61/87; 41/59	9/7; 56/44	0,84
Diagnosis (n; %)			-
BA	86; 58		
Byler's disease	15; 10		
BH	14; 10		
Allagille syndrome	12; 8		
Caroli disease	5; 3		
Other	16; 11		

Table 1.
Baseline patients' characteristics, BA, biliary atresia; BH, biliary hypoplasia.

and sex of the children included in the study and those of healthy children did not differ (p = 0.78 and p = 0.84, respectively).

All recipients underwent hepatectomy with preservation of the inferior vena cava, followed by implantation of a liver fragment (left lateral sector) in an orthotopic position. Blood flow characteristics were monitored intraoperatively using ultrasound. Bile duct reconstruction was performed after graft revascularization on the defunctionalized jejunal loop.

5.2 Methods

Patients were examined before and after LT in accordance with the protocol approved at Shumakov Center. Routine examination and treatment of patients before and after LT were carried out in accordance with the clinical guidelines of the Russian Transplant Society and international consensus guidelines.

The required methods were:

1. Physical examination, including observation, palpation, percussion, and auscultation;

2. Laboratory examination: biochemical blood test, coagulogram, general blood test, hemodynamic parameters.

3. Instrumental methods: ultrasound, echocardiography and the study of the electrical activity of the heart; multislice CT scan of the brain, chest, and abdominal cavity with intravenous contrast.

Double- and triple-combination immunosuppressive therapies were used at Shumakov Center.

During preparation for donor organ implantation, all recipients received 10 mg of basiliximab. Reintroduction of basiliximab was performed on day 4 after surgery. Immediately before graft reperfusion, methylprednisolone was administered at a dose of 10 mg/kg.

Maintenance immunosuppressive therapy included tacrolimus (at 6–8 ng/mL target concentration), glucocorticoids, and mycophenolic acid preparations. Cyclosporin A was administered in a few cases with tacrolimus neurotoxicity. Gradual reduction of methylprednisolone doses to 4–2 mg per day was continued.

Mycophenolic acid preparations were administered as the third component of the therapy in the absence of haplotype compatibility with donors and when a rejection reaction occurred.

In the case of rejection, intravenous pulse therapy was performed with high doses of glucocorticoids (20 mg/kg).

The concentration of biomarkers was studied in plasma or serum obtained from peripheral blood after centrifugation at 1500–2500 g at room temperature. Blood was collected in plastic tubes (BD Vacutainer, Becton Dickinson, USA) without anticoagulant or with sodium citrate as an anticoagulant. Serum and plasma samples were stored at −50°C until analysis.

GH levels were measured by enzyme immunoassay using a reagent kit (DBC, Canada) according to the manufacturer's instructions in blood serum. The optical density of the reaction mixture was measured at 450 nm wavelength using a spectrophotometer. Analytical sensitivity of the method was 0.2 ng/ml. The range of determined concentrations is 0.2–50 ng/ml.

IGF-1 levels were measured by the sandwich immunoassay method using polyclonal sheep antibodies and monoclonal antibodies against IGF-1 in plasma (IDS Ltd. 77 OCTEIA® IGF-1, UK). The analytical sensitivity of the method was 3.1 ng/ml. The range of determined concentrations is 12–627 ng/ml.

5.3 Statistical analysis

Data are presented as mean and standard deviation (M ± SD), upper and lower limits of the 95% confidence interval for parametric variables, and median and interquartile range from 25th to 75th percentile for nonparametric variables. The data were statistically processed using nonparametric statistics methods: paired Wilcoxon test was calculated to compare dependent samples, and the Mann-Whitney or Kruskal-Wallis U test was used to compare independent variables. Spearman rank correlation coefficient was calculated to evaluate the relationship between quantitative and qualitative ordinal signs. In the case of error probability ($p < 0.05$), the differences were considered statistically significant. The obtained data were processed using Microsoft Office Excel with the IBM SPSS STATISTICS 20 software package for scientific and technical calculations (IBM SPSS Inc., USA).

6. GH/IGF-1 and anthropometric characteristics of children with cholestatic diseases

Children suffering from congenital hepatobiliary diseases have multiple and various homeostasis disorders, largely due to a violation of the key functions of the liver—synthesizing, detoxifying, regulatory, etc.—which ultimately lead to growth retardation and underweight.

In the patients before LT, the average body height was 71.2 ± 8.2 cm, and body mass was 7.9 ± 2.3 kg. Comparison of average height and weight in the study group with reference values for healthy children of the same age (mean age, 11 months) showed that the average height of the recipients was significantly lower than the average reference value for healthy children according to World Health Organization (WHO) [55]—75 ± 6 cm, p = 0.00, and the average body mass of the recipients was significantly lower than reference values - 9.5 ± 2 kg, p = 0.00.

It should be noted that a correct comparison of anthropometric indicators of the study group with healthy patients is a difficult task due to the absence of a group of healthy children of the same ethnic group that is similar in age and sex. Therefore, the authors consider it possible to compare the histogram of the distribution of a trait with a normal distribution curve superimposed on a similar range of an anthropometric indicator (**Figure 5**).

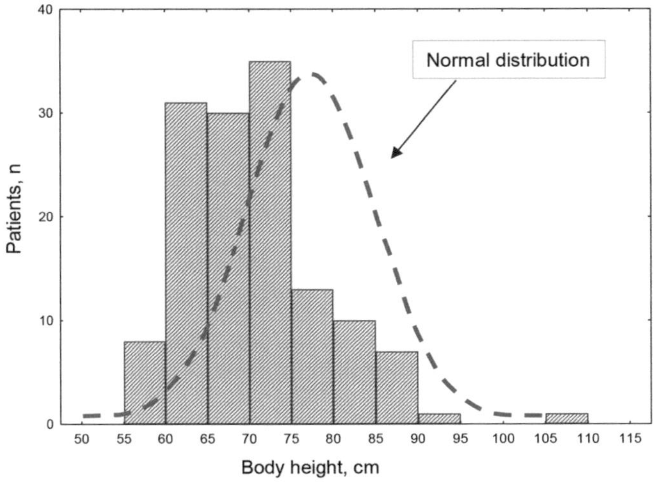

A

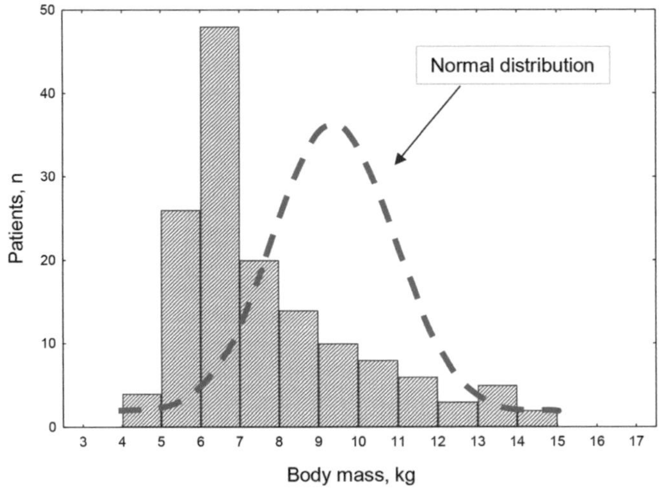

B

Figure 5.
Histogram of the distribution of body height (A) and body mass (B) of children with ESLD in comparison with the normal distribution curve.

The figure shows that the histograms of body height and body mass distribution of in children with ESLD differ significantly from the normal distribution curves and are shifted to lower values.

Thus, in the examined group of children with ESLD, anthropometric indicators are lower than the values typical for healthy children of the same age. Obviously, the low body height and body mass of patients are associated with ESLD.

The Pediatric End-stage Liver Disease (PELD) scale was used to assess the relative severity of liver disease. This scale is used to evaluate the probable 90-day survival of patients awaiting LT and was developed to objectively prioritize LT candidates in children. Like the Model End-stage Liver Disease (MELD) score, the PELD score was derived from biological markers of liver function (albumin, bilirubin, and international normalized prothrombin time ratio) and growth retardation [56, 57].

The PELD score, which reflects the severity of liver disease, averaged 19.0 ± 8.2 (median, 18.0; range, 0 to 51) in the study group (**Figure 6**).

The distribution histogram of the PELD score in the examined children shows that in most patients, the score ranges from 10 to 25 and indicates that they have severe liver failure.

To assess the relationship between the severity of liver failure and the anthropometric parameters of the liver recipients, a correlation analysis was carried out between the PELD scores and the body height or body mass of the patients (**Figure 7**).

The dependencies presented in the graphs indicate that patients' height and weight, determined before transplantation, significantly correlate with the PELD score. The higher the PELD score in patients before radical treatment of the underlying disease—liver fragment transplantation—the lower the level of anthropometric parameters (body mass and height).

The obtained result suggests that, along with other changes, there may be impairment of hormonal growth regulation in this group of children. The most important growth regulators are GH produced by the pituitary gland and IGF-1 synthesized in

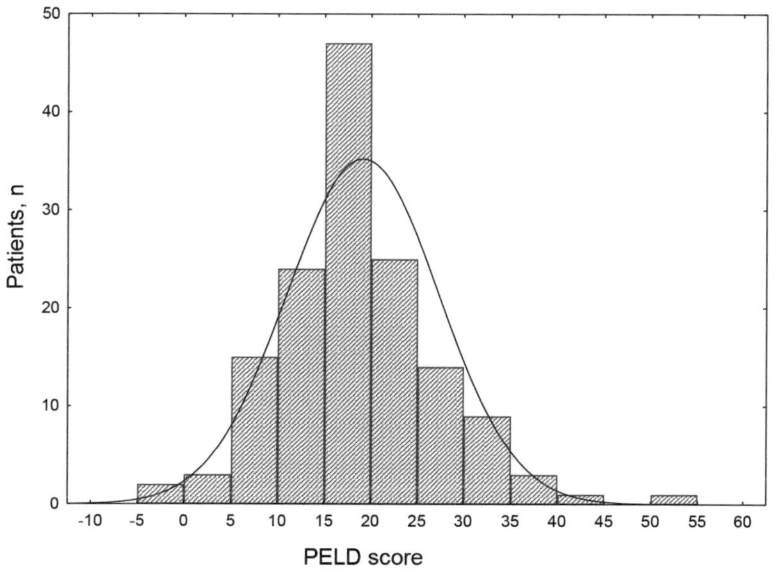

Figure 6.
Histogram of the distribution of PELD score in children with ESLD.

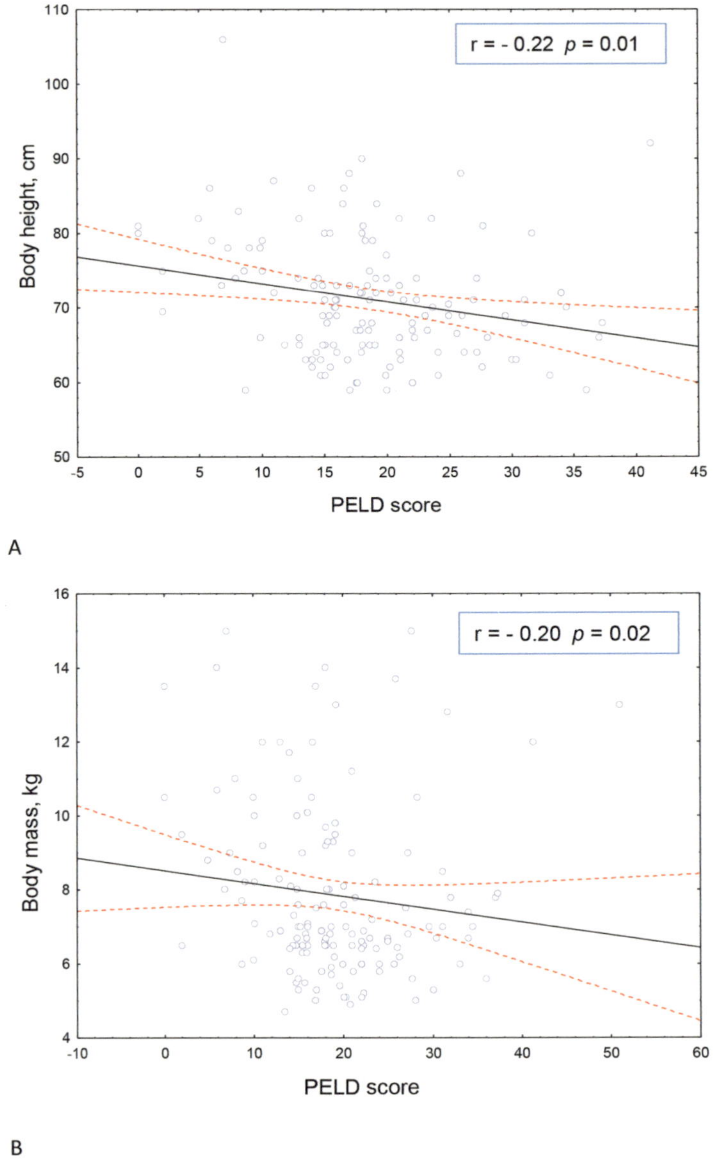

Figure 7.
Correlation of PELD score with body height (A) and body mass (B) of pediatric liver recipients.

the liver. Comparative analysis of GH levels in patients before LT and healthy children of the same age indicates a significant difference in the indicator (**Figure 8**).

The median plasma GH level of children with ESLD was 4.3 (1.6–7.2) ng/ml and was significantly higher than in healthy children of the same age—1.2 (0.3–2.4) ng/ml, $p = 0.001$. Thus, there is a paradoxical situation when high GH levels in children with ESLD are combined with growth retardation and weight loss.

A comparative analysis of IGF-1 levels in patients before LT and healthy children of the same age also showed significant differences (**Figure 9**).

The median IGF-1 level in children with liver failure was 7.9 (0.0–25.1) ng/ml and was significantly lower than in healthy children: 38.0 (26.9–56.5) ng/ml, $p = 0.001$;

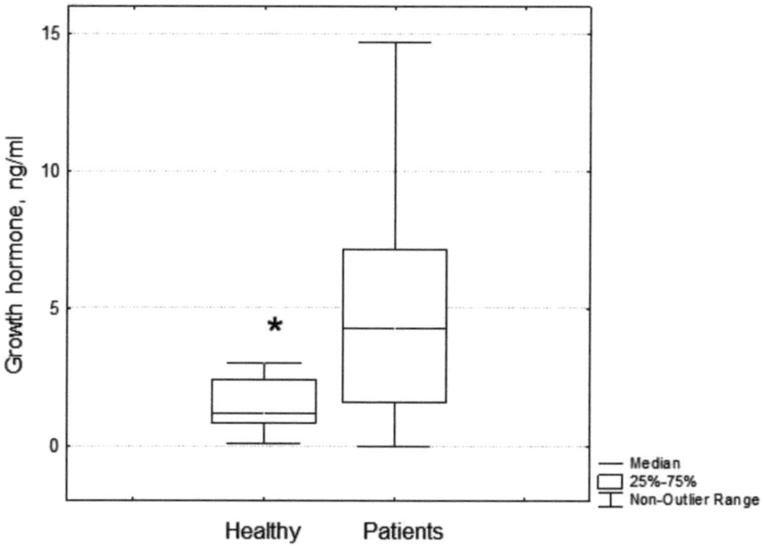

Figure 8.
*The level of GH in the blood of healthy children and patients with ESLD, * - p < 0.05.*

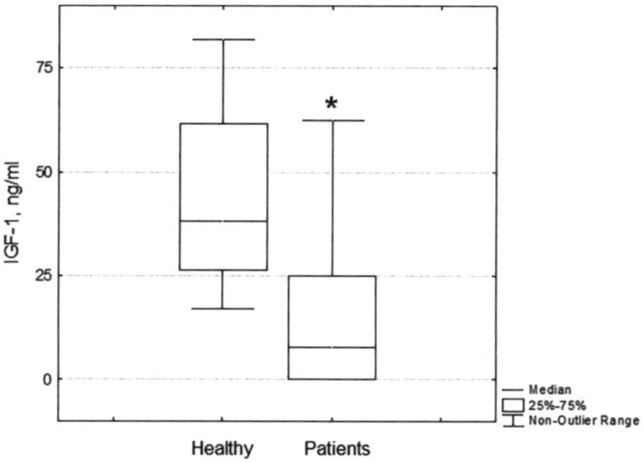

Figure 9.
*The level of IGF-1 in the blood of healthy children and patients with ESLD, * - p < 0.05.*

this is consistent with the literature data and is associated with the inability of the damaged liver to produce this growth regulator [58, 59]. Significantly low level of IGF-1 in liver recipients, obviously, reflects impaired synthesis and production of IGF-1 in the liver and explains the high level of GH in the blood.

Correlation analysis between GH and IGF-1 levels in children before transplantation revealed no significant relationship between hormone levels (**Figure 10**).

The analysis confirms that in children with ESLD, there is a violation of the relationship between GH and IGF-1 concentrations in the blood.

To characterize the hormonal regulation of anthropometric parameters and liver function, we analyzed the correlations between the GH and IGF-1 levels in children with height, body mass, and PELD scores (**Table 2**).

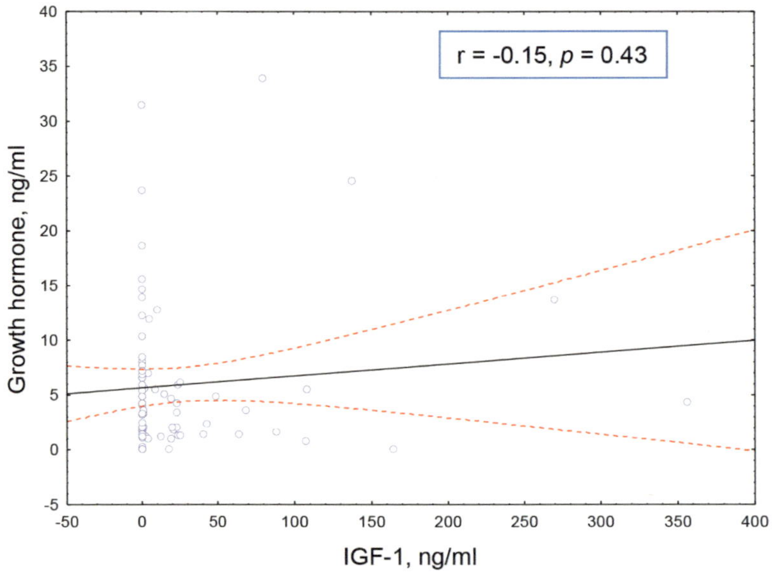

Figure 10.
Correlation between IGF-1 and GH blood plasma levels in children with ESLD.

Biomarker	Correlation coefficient		
	Body mass	**Body height**	**PELD**
GH	−0.17	−0.14	0.02
IGF-1	0.29*	0.28*	0.15
*—$p < 0.05$.			

Table 2.
Correlation of levels of GH (GH) and IGF-1 in the blood plasma of recipients with anthropometric parameters and the value of the PELD index before LT.

The analysis revealed no statistically significant associations between GH levels and body height, body mass, or PELD score. Whereas IGF-1 concentrations in the blood were significantly associated with body height and body mass ($p = 0.001$), but not with PELD score.

The result obtained shows that in children with ESLD, their body height and body mass are not associated with plasma GH levels, while IGF-1 levels correlate with these anthropometric characteristics. In the study group, no correlation between hormone levels and ESLD was found, which is inconsistent with an earlier study that showed a relationship between GH levels and PELD score [60, 61]. This can be explained by many factors, in particular, differences in the structure of the incidence and degree of liver damage in patients in the studied samples.

7. GH/IGF-1 and clinical parameters of children with cholestatic diseases before LT

The patients included in the study had various etiologies of ESLD. Analysis of the structure of liver diseases showed that the largest group, about 60%, consisted of recipients with BA, a congenital malformation of the biliary tract (**Figure 11**).

Plasma GH levels in the patients differed depending on the etiology of liver disease: 5.5 (4.8–13.7) ng/ml in Alagille syndrome, 4.9 (2.0–7.3) ng/ml in BA, 1.6 (1.2–2.3) ng/ml in others, 3.9 (1.3–4.8) ng/ml in Byler disease, 3.6 (1.5–5.2) ng/ml in BH, 3.9 (0.0–7.7) ng/ml in Caroli disease (**Figure 12**).

Comparative analysis of GH levels in children with different etiologies of liver disease using the Kruskal-Wallis test for several independent variables revealed no statistically significant differences in GH concentrations (p = 0.16). Pairwise comparison of GH concentrations in the groups with different diagnoses using the

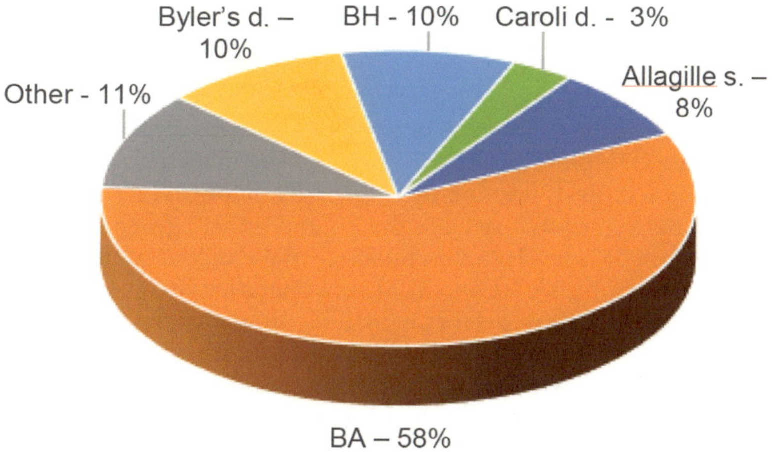

Figure 11.
Structure of morbidity in the studied liver transplant recipients, BA, biliary atresia; BH, biliary hypoplasia.

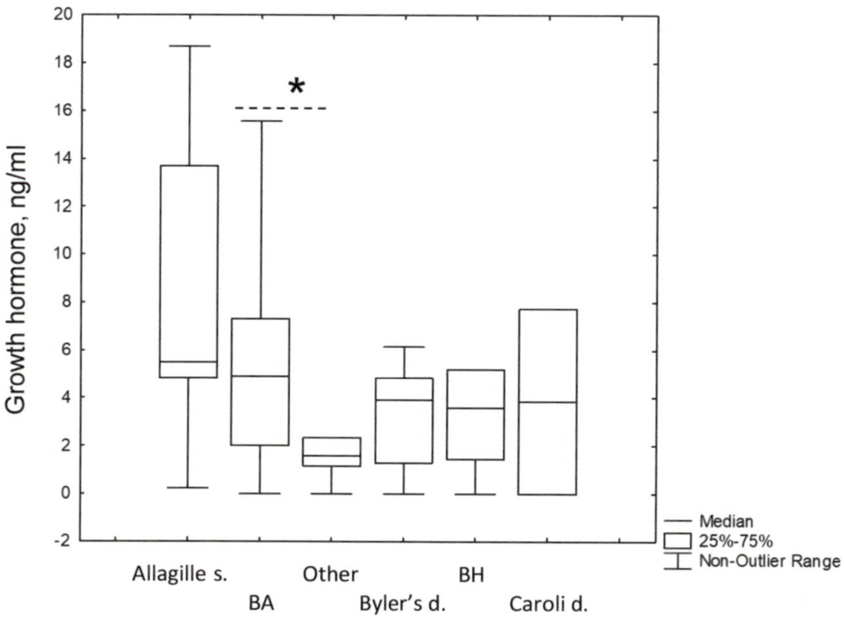

Figure 12.
*GH levels in the blood of children with various etiologies of liver disease, * - p = 0.012 compared with the level in children with BA, BA, biliary atresia; BH, biliary hypoplasia.*

Mann-Whitney U test showed that the plasma GH level in children with ESLD resulting from BA was significantly higher than in patients grouped under "other" diseases (autoimmune and fulminant hepatitis, Von Gierke disease, portal vein developmental anomaly, Crigler-Najjar syndrome, Budd-Chiari syndrome, hepatoblastoma, alpha-1 antitrypsin deficiency, and previous liver transplant dysfunction), $p = 0.012$. No significant differences were found between the other groups.

Analysis of plasma IGF-1 concentrations depending on the etiology of liver disease showed that the hormone content in the blood of patients with Alagille syndrome was 12.6 (0.0–28.1) ng/ml, with BA 1.2 (0.0–20.0) ng/ml, others 42.3(3.3–72.3) ng/ml, Byler disease 25.1 (14.7–51.7) ng/ml, BH 1.1 (0.0–12.0) ng/ml, Caroli disease 3.6(0.0–8.7) ng/ml (**Figure 13**).

Multiple comparative analysis showed significant differences in IGF-1 levels in children with different etiologies of liver disease, $p = 0.01$ according to Kruskal-Wallis test. Pairwise comparison of IGF-1 levels in the groups with different diagnoses using the Mann-Whitney U test revealed that IGF-1 levels significantly differed in the group BA and "others" ($p = 0.004$), BA, and Byler disease groups (0.012). Differences were also found in the hormone level between the BH and "other" groups ($p = 0.013$).

It is possible that IGF-1 levels in the children depend on the etiology of liver disease; however, since the "other" group included various pathologies and each of them is small, further studies are needed for an unambiguous conclusion.

It is known that the nature of GH secretion differs in males and females and depends on age [62]. In our study, we could not identify a significant correlation between GH level and the age of patients before transplantation ($r = -0.18$, $p = 0.10$). This could be probably due to both impaired hormone secretion associated with the disease and by the age structure of the sample used in our work. In the control group,

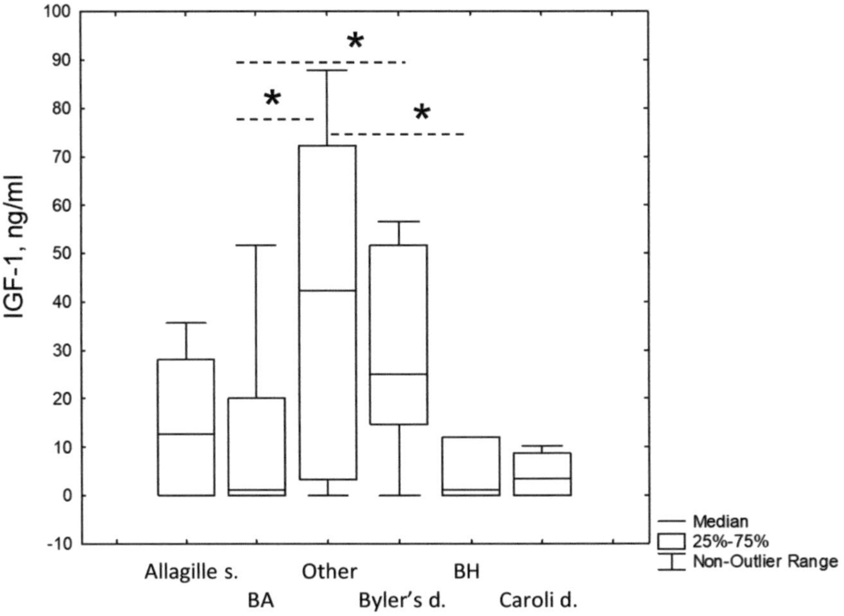

Figure 13.
*The level of IGF-1 in the blood of children with various etiologies of liver disease, * - p < 0.05 in comparison with the level in children with BA; BA, biliary atresia; BH, biliary hypoplasia.*

there was also no significant correlation with age, which can be explained by the narrow age range of the sample and/or a small number of observations.

A comparative analysis of plasma GH levels in girls (4.7 (1.7–7.1) ng/ml) and boys (3.4 (1.6–7.2) ng/ml) with ESLD revealed no significant differences, $p = 0.41$. In the control group, we could not also identify significant differences in GH levels in healthy boys and girls.

Serum IGF-1 levels, like GH, change with age, while there are practically no sex differences in protein levels. In the studied children, no significant correlation was found between IGF-1 levels and age (r = −0.001, $p = 0.96$). In the control group, no significant differences were also found, which is obviously due to the narrow age range in the groups included in the study.

The IGF-1 content in girls and boys with ESLD was 3.9 (0.0–23.5) ng/ml and 11.1 (0.0–33.6) ng/ml, respectively, and did not differ significantly, $p = 0.38$. The result obtained showed that IGF-1 concentration in children with ESLD is not associated with the sex of the child. No significant differences were found in the control group either.

The absence of age and sex differences in GH and IGF-1 levels in the group of children aged 3 months to 5 years with ESLD can be explained by a number of reasons—the absence of such differences at this early age, and impaired hormone secretion associated with the disease. Our data do not provide an unambiguous answer to this question due to the age composition of the study group. Of course, studies with a large number of observations and in more homogeneous age groups will be informative.

8. Changes in the content of GH and IGF-1 in the blood plasma of children after LT

To assess the effect of LT on the levels of the studied hormones in the early and late post-transplant period, the levels in the recipients were measured 1 month and 1 year after the transplantation.

Significant changes in the plasma GH levels were found in the children before, a month and a year after liver transplantation (**Figure 14**).

GH level 1 month after LT was 1.4 (1.1–2.4) ng/ml. It was significantly lower before the operation ($p = 0.001$), but did not differ from the hormone level in healthy children ($p = 0.74$).

A year after transplantation, the GH level was 2.5 (1.5–5.7) ng/ml, which was also significantly lower than before transplantation (p = 0.049) and did not differ from the healthy children's level ($p = 0.67$).

The change in IGF-1 concentrations in the children a month and a year after liver fragment transplantation is shown in **Figure 15**.

In the blood plasma of the pediatric recipients, there was a significant increase in IGF-1 levels compared with the values before transplantation. One month later, the IGF-1 concentration was 73.5 (52.1–128.5) ng/ml ($p = 0.000$ compared with the concentration before transplantation). One year after transplantation, plasma IGF-1 level was 76.1 (42.9–111.7) ng/ml and was also significantly higher than before liver transplantation ($p = 0.000$ compared with the level before transplantation) and did not differ from the level in healthy children ($p = 0.48$).

A comparative analysis of changes in levels of the hormones in the pediatric recipients before and after LT depending on the gender of the patient was carried out. It was revealed that GH concentrations significantly differ among girls and boys after LT (**Figure 16**).

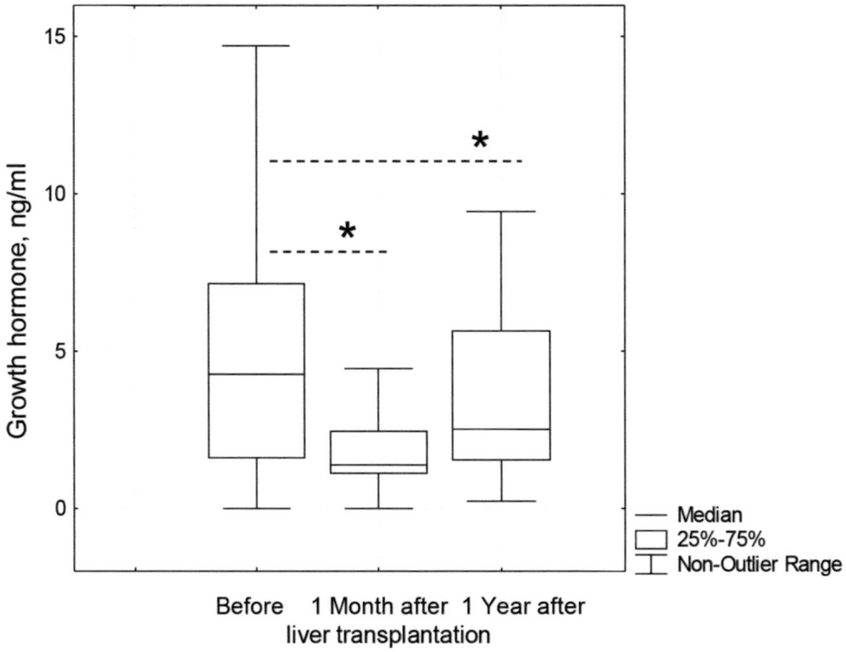

Figure 14.
*The level of GH in the blood of children after LT. * – p < 0.05 in comparison with the level before LT.*

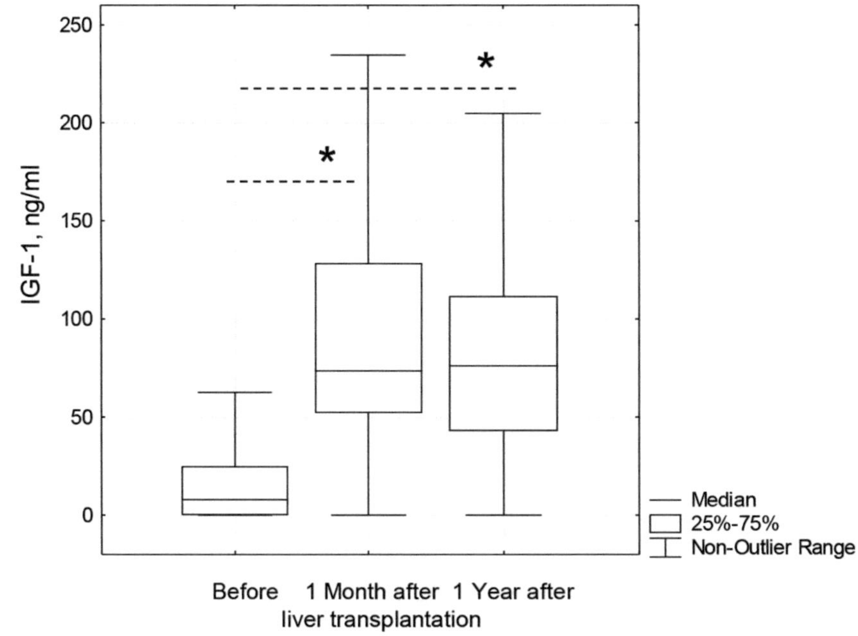

Figure 15.
*The level of IGF-1 in the blood of children after LT, * - p < 0.05 in comparison with the level before LT.*

The analysis showed that plasma GH concentrations in girls and boys did not differ before LT - 4.7 (1.7–7.1) ng/ml and 3.4 (1.6–7.2) ng/ml, respectively, $p = 0.41$, and a month after - 1.4 (1.1–2.5) ng/ml and 1.6 (1.2–2.4) ng/ml, respectively, $p = 0.8$.

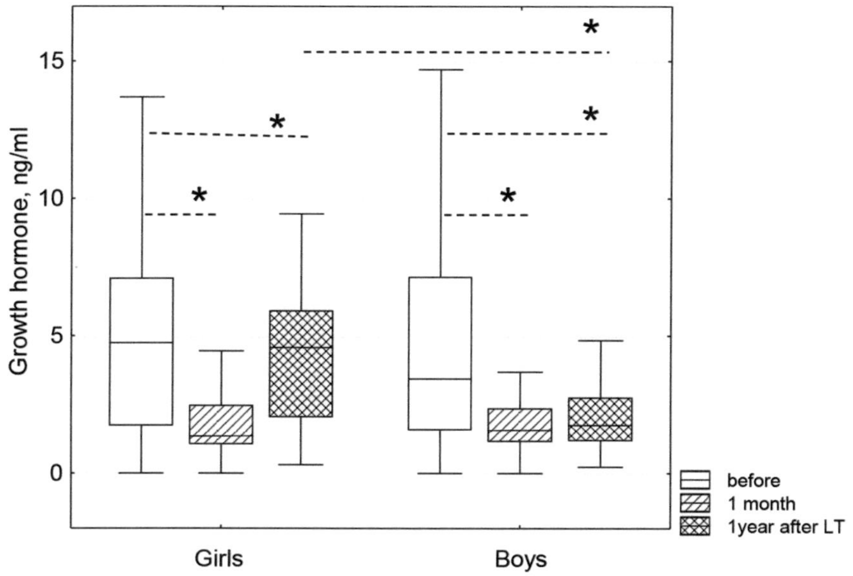

Figure 16.
*The content of GH in the blood of children after LT, depending on gender, * - p < 0.05.*

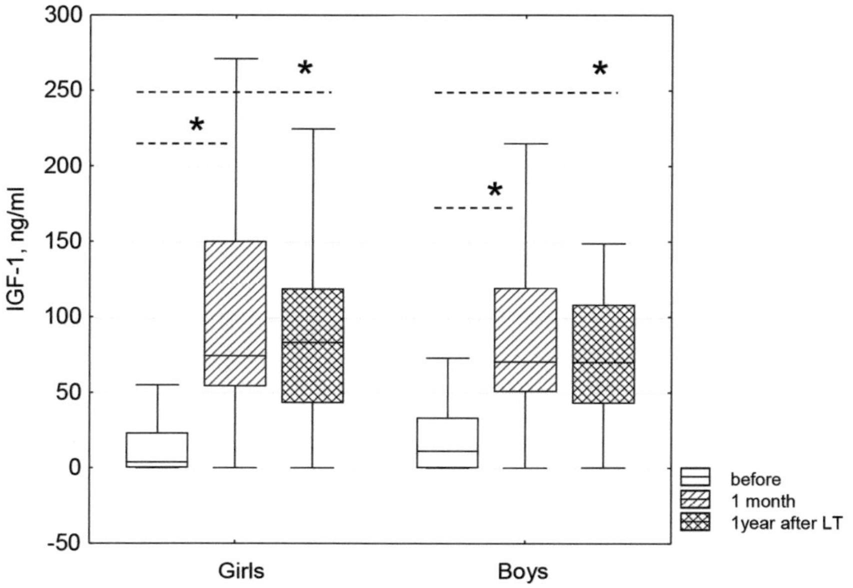

Figure 17.
*The content of IGF-1 in the blood of children after LT, depending on gender, * - p < 0.05.*

Whereas a year after transplantation, the levels in girls were significantly higher than those in boys—4.6 (2.1–5.9) ng/ml and 1.8 (1.2–2.8) ng/ml, respectively, $p = 0.002$. It is difficult to assess the reasons for the differences in GH levels after LT in boys and girls in the study sample, because for healthy children under 6 years of age, according to the reference values, differences in GH levels between the sexes are insignificant. However, given the sex differences in GH hormonal regulation in later puberty, it cannot be ruled out that periods of differences between boys and girls can be observed in young children. Prescription of immunosuppressants after LT can also affect the result.

After LT, IGF-1 levels increased in both girls and boys equally (**Figure 17**).

There were no differences in IGF-1 levels in the children depending on gender, as before transplantation: IGF-1 levels in girls and boys were 3.9 (0.0–23.5) ng/ml and 11.1 (0.0–33.6) ng/ml, respectively, $p = 0.38$, and a month after it - 74.2 (54.0–150.3) ng/ml and 70.4 (50.6–119.5) ng/ml, respectively, $p = 0.37$. A year after the operation, the IGF-1 level in girls did not significantly differ from that of boys—83.1 (43.0–118.9) ng/ml and 70.1 (42.9–108.4) ng/ml, respectively, $p = 0.68$.

9. Relationship between GH/IGF-1 levels and anthropometric characteristics of recipients after LT

Analysis of changes in the anthropometric parameters of the pediatric recipients was carried out 1 year after LT, because 1 month after the operation, these indicators practically do not differ from preoperative ones.

One year after transplantation, the average height (82.1 ± 7.6 cm) and body mass (11.5 ± 2.2 kg) in the examined group were significantly higher than before LT ($p = 0.00$ for both parameters). **Figure 18** shows the change in the histogram of the distribution of height and body mass of the examined children one year after LT.

The results presented in the histogram show that the children's height and weight change a year after LT, close to the normal distribution curve.

Comparison of the average height and body mass in the study group with the reference values for healthy children of the same age (mean age, 23 months) showed that the body mass of the recipients did not significantly differ from the reference values (12 ± 2 kg), p = 0.06.

The average height of the recipients was still significantly below the average reference value for healthy children according to WHO standards (87 ± 7 cm), $p = 0.00$ [55].

Currently, there are no published studies of long-term follow-up of the anthropometric parameters of liver transplant pediatric recipients, and therefore, it is difficult to assess how fully their growth is restored in the long-term post-transplant period. For example, it has been found that for child kidney recipients, there is a problem of growth retardation both before and after transplantation, which is associated with many reasons, including taking immunosuppressants. For such patients, recommendations have been developed on how to correct growth retardation by taking recombinant GH [63].

As already noted, we compared the anthropometric indicators of the study group with healthy children by comparing the histogram of the distribution of height and body mass with a normal distribution curve superimposed on a similar range of the anthropometric indicator (**Figure 19**).

An analysis of the relationship between GH/IGF-1 levels and anthropometric parameters 1 year after LT showed that GH levels significantly correlate with body mass ($r = -0.27$, $p = 0.046$)—the lower the plasma GH level, the greater the body mass. The correlation of GH levels with patients' height was also negative, but did not reach statistical significance ($r = -0.20$, $p = 0.143$). The correlation of IGF-1 content in the blood with the height ($r = 0.19$, $p = 0.055$) and body mass of patients ($r = 0.18$,

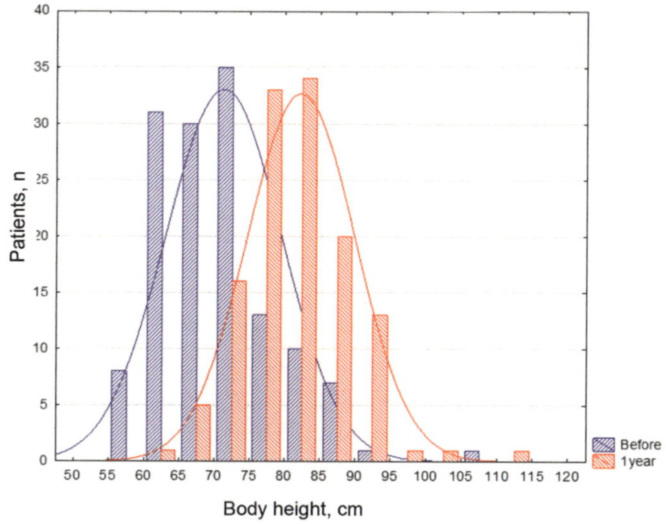

A

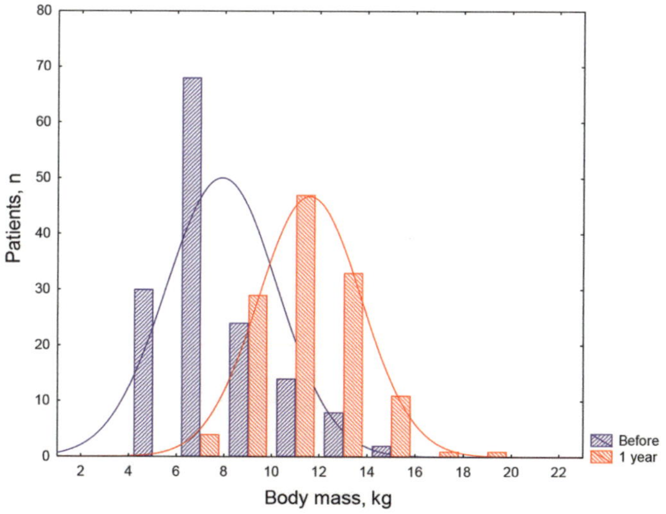

B

Figure 18.
Change in the histogram of the distribution of height (A) and body mass (B) of recipient children 1 year after LT.

p = 0.074) was positive, but did not reach statistical significance. The described relationships reflect the restoration of the physiological regulation of anthropometric parameters in children after LT due to the restoration of IGF-1 production by the graft and provision of GH bioavailability.

A correlation analysis was carried out to assess the relationship between GH and IGF-1 levels in the blood of children after LT. It revealed the restoration of the relationship between the levels of hormones in pediatric recipients already a month after

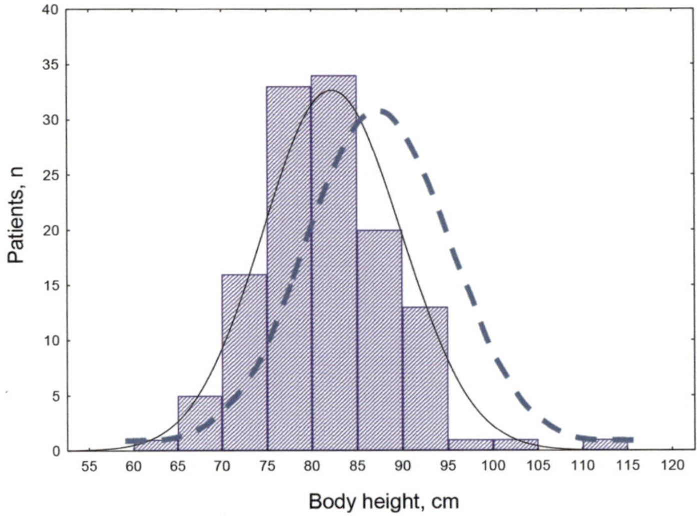

A

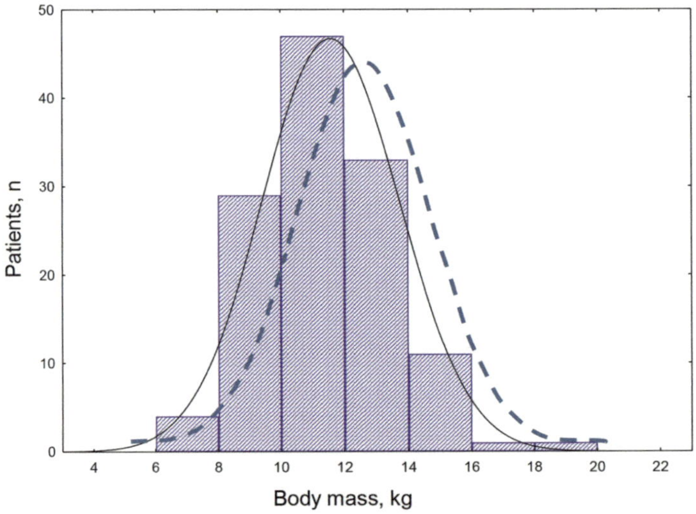

B

Figure 19.
Histograms of the distribution of height (A) and body mass (B) of recipient children one year after LT in comparison with the normal distribution curve.

LT ($r = -0.28$, $p = 0.011$); the correlation strength increased a year after LT: $r = -0.47$, $p = 0.01$ (**Figure 20**).

The presented results show that in pediatric recipients a year after LT, a statistically significant negative correlation is observed between the plasma GH and IGF-1 levels, which indicates normalization of the relationship between the hormones and normalization of the regulatory mechanisms of anthropometric characteristics in the children. The IGF-1-to-GH ratio can serve as a kind of integral indicator of such normalization.

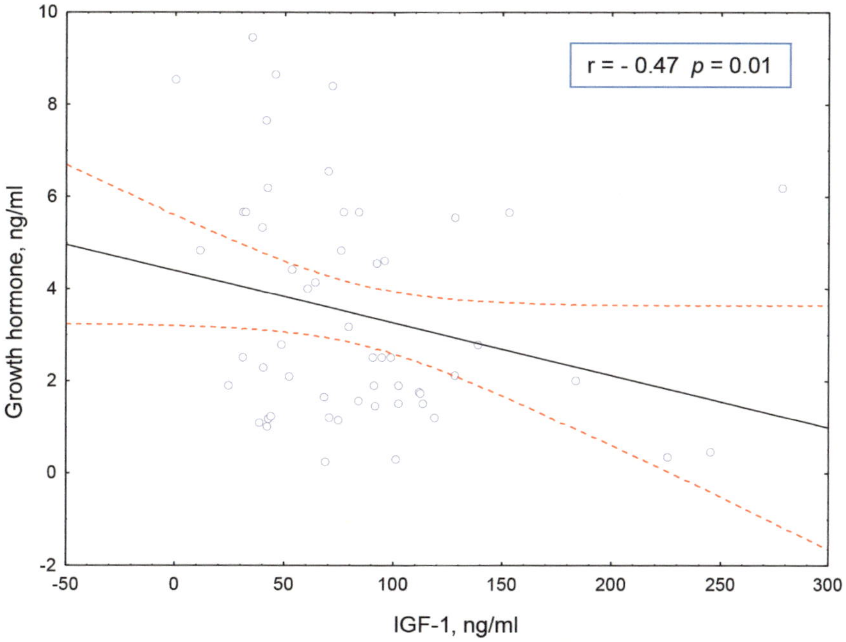

Figure 20.
Correlation of IGF-1 and GH levels in children 1 year after LT.

Analysis before and after LT showed that this ratio increased significantly after LT (**Figure 21**).

Before LT, the IGF-1-to-GH ratio was 5.5 ± 4.2 and was significantly lower than that in healthy children - 37.1 ± 18.4 (p = 0.001); a month later, this parameter had increased by almost an order of magnitude, reaching 62.2 ± 41.0 (p = 0.001), and a year later, it had decreased to 32.6 ± 21.8 and did not differ from that in healthy children (p = 0.98), but was higher than before transplantation (p = 0.002). The data obtained suggest that mutual regulation of the GR/IGF-1 axis is restored 1 year after LT.

Obviously, normalization of the feedback between the GH and IGF-1 levels a year after LT is an indicator of the restoration of the normal ratio of the components of the central and peripheral growth regulation system and is a consequence of the restoration of liver function.

Plasma GH and IGF-1 levels not only depend on liver function, but also largely determine its condition and can be considered as indicators of liver function in patients with hepatobiliary diseases. Changes in these levels after LT can serve as an objective indicator of the degree of normalization of the synthetic function of the graft and restoration of neurohumoral regulation of the body of pediatric liver recipients.

Separate long-term post-transplant follow-up of pediatric liver recipients shows that with good graft function, the anthropometric characteristics of children can be fully restored and do not differ from those of healthy children of the same age.

Figure 22 shows a photo of patient P. (in the center outlined by a dotted line) with her classmates 7 years after LT. It can be seen that her height is practically the same as that of her peers.

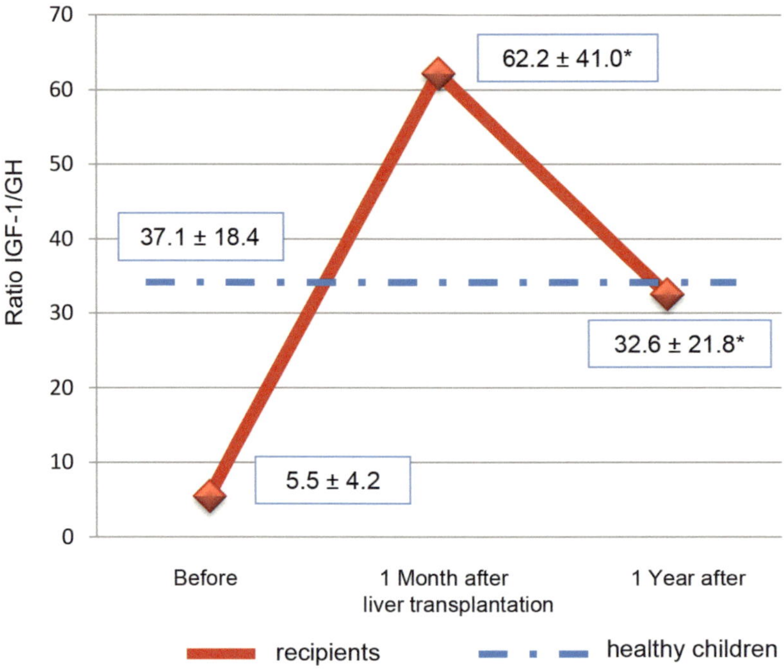

Figure 21.
The ratio of IGF-1/GH levels in pediatric recipients before, 1 month and 1 year after LT.

Figure 22.
Patient P., 7 years after LT.

This study has revealed that the GH/IGF-1 axis plays a critical role in the pathogenesis of ESLD in pediatric patients. Children with ESLD have high PELD score, growth retardation, and decreased body mass. Their serum GH levels are much higher than in healthy children, while IGF-1 levels are lower than in healthy children. After LT, the GH/IGF-1 axis recovers, improving anthropometric parameters in pediatric recipients. Result of the study suggests that the GH/IGF-1 axis is of clinical significance in pediatric liver recipients. The GH/IGF-1 levels and the IGF-1-to-GH ratio can reflect liver graft function.

10. Conclusion

ESLD results in a decline in liver production and deficiency of IGF-1, which impair GH bioavailability for peripheral tissues. This hormonal disorder leads to delayed growth in children with congenital cholestatic diseases.

Our studies have shown that children with ESLD have high GH and low IGF-1 levels, which is combined with growth failure and body mass deficit.

After pediatric LT, GH concentrations fall, while those of IGF-1 increase. As early as a month after LT, the plasma concentrations of both hormones do not differ from those in healthy children, and a year after the transplantation, they remain at normal levels.

After LT, the relationship between plasma GH and IGF-1 levels is restored, which was disrupted in children with ESLD.

Normalization of the ratio between serum levels of IGF-1 and GH a year after LT can serve as an integral indicator of the restoration of hormonal regulation of the physical development of pediatric recipients. A year after LT, GH and IGF-1 levels in pediatric recipients are comparable to those in healthy children.

GH bioavailability, mediated through IGF-1, is restored due to normal production of IGF-1 by the graft. The body height and body mass of pediatric recipients are also comparable to those of healthy children. This reflects the restoration of central and peripheral hormonal regulation of growth, which is a consequence of graft functioning.

Acknowledgements

The authors would like to thank all the transplant teams from Shumakov Center, who provided patients' data and samples for this study.

The study was supported by research grants from the Ministry of Health of the Russian Federation.

Conflict of interest

The authors declare no conflict of interest.

Author details

Sergey V. Gautier[1,2], Olga P. Shevchenko[1,2], Olga M. Tsirulnikova[2], Rivada M. Kurabekova[1*] and Irina E. Pashkova[1]

1 Shumakov National Medical Research Center of Transplantology and Artificial Organs of the Ministry of Healthcare of the Russian Federation, Moscow, Russian Federation

2 I.M. Sechenov First Moscow State Medical University of the Ministry of Healthcare of the Russian Federation, Moscow, Russian Federation

*Address all correspondence to: kourabr@yandex.ru

References

[1] Meirelles Junior RF, Salvalaggio P, Rezende MB, Evangelista AS, Guardia BD, et al. Liver transplantation: History, outcomes and perspectives. Einstein. 2015;**13**(1):149-152. DOI: 10.1590/S1679-45082015RW3164

[2] Starzl TE, Marchioro TL, Vonkaulla KN, Hermann G, Brittain RS, Waddell WR. Homotransplantation of the liver in humans. Surgery, Gynecology & Obstetrics. 1963;**117**:659-676

[3] StatistaGmbH. Estimated number of worldwide liver transplants in 2020, by region. 2022. Available from: https://www.statista.com/statistics/398685/liver-transplants-by-world-region/. [Accessed: 2022-08-15]

[4] Berryman DE, Christiansen JS, Johannsson G, Thorner MO, Kopchick JJ. Role of the GH/IGF-1 axis in lifespan and healthspan: Lessons from animal models. Growth Hormone & IGF Research. 2008;**18**(6):455-471. DOI: 10.1016/j.ghir.2008.05.005

[5] Fujio A, Kawagishi N, Echizenya T, Tokodai K, Nakanishi C, Miyagi S, et al. Long-term survival with growth hormone replacement after liver transplantation of pediatric nonalcoholic steatohepatitis complicating acquired hypopituitarism. The Tohoku Journal of Experimental Medicine. 2015;**235**(1):61-67. DOI: 10.1620/tjem.235.61

[6] Nicolini D, Mocchegiani F, Palmonella G, Coletta M, Brugia M, Montalti R, et al. Postoperative insulin-like growth factor 1 levels reflect the graft's function and predict survival after liver transplantation. PLoS One. [serial on the Internet]. 2015;**10**(7):e0133153. DOI: 10.1371/journal.pone.0133153

[7] Murray PG, Clayton PE. Disorders of Growth Hormone in Childhood. South Dartmouth (MA): MDText.com, Inc.; 2000. p. 2022

[8] Maes M, Sokal E, Otte JB. Growth factors in children with end-stage liver disease before and after liver transplantation: A review. Pediatric Transplantation. 1997;**1**(2):171-175

[9] Rodeck B, Kardorff R, Melter M, Ehrich JH. Improvement of growth after growth hormone treatment in children who undergo liver transplantation. Journal of Pediatric Gastroenterology and Nutrition. 2000;**31**(3):286-290. DOI:10.1097/00005176-200009000-00016

[10] Weigent DA. Lymphocyte GH-axis hormones in immunity. Cellular Immunology. 2013;**285**(1-2):118-132. DOI: 10.1016/j.cellimm.2013.10.003

[11] Conover CA, Oxvig C. PAPP-A: A promising therapeutic target for healthy longevity. Aging Cell. 2017;**16**(2):205-209. DOI: 10.1111/acel.12564

[12] Hawkes CP, Grimberg A. Measuring growth hormone and insulin-like growth factor-I in infants: What is normal? Pediatric Endocrinology Reviews. 2013;**11**(2):126-146

[13] Katsumata N. Standardization of growth hormone and insulin-like growth factor-I measurements. Pediatric Endocrinology Reviews. 2017;**1**(Suppl. 1):209-215. DOI: 10.17458/per.vol14.2017.k.standardizationgrowth

[14] Leung KC, Ho KK. Measurement of growth hormone, insulin-like growth factor I and their binding proteins: The clinical aspects. Clinica Chimica Acta. 2001;**313**(1-2):119-123

[15] Junnila RK, List EO, Berryman DE, Murrey JW, Kopchick JJ. The GH/IGF-1 axis in ageing and longevity. Nature Reviews Endocrinology. 2013;**9**(6):366-376. DOI: 10.1038/nrendo.2013.67

[16] Cignarelli A, Genchi VA, Le Grazie G, Caruso I, Marrano N, Biondi G, et al. Mini review: Effect of GLP-1 receptor agonists and SGLT-2 inhibitors on the growth hormone/IGF axis. Frontiers in Endocrinology. 2022;**13**:846903. DOI: 10.3389/fendo.2022.846903

[17] De Palo EF, Bassanello M, Lancerin F, Spinella P, Gatti R, D'Amico D, et al. GH/IGF system, cirrhosis and liver transplantation. Clinica Chimica Acta. 2001;**310**(1):31-37

[18] Vázquez-Borrego MC, Del Rio-Moreno M, Kineman RD. Towards understanding the direct and indirect actions of growth hormone in controlling hepatocyte carbohydrate and lipid metabolism. Cell. 2021;**10**(10):2532. DOI: 10.3390/cells10102532

[19] Hu M, Wang C, Zhang GY, Saito M, Wang YM, Fernandez MA, et al. Infiltrating Foxp3(+) regulatory T cells from spontaneously tolerant kidney allografts demonstrate donor-specific tolerance. American Journal of Transplantation. 2013;**13**(11):2819-2830. DOI: 10.1111/ajt.12445

[20] Jiang Z, Chen Y, Feng X, Jiang J, Chen T, Xie H, et al. Hepatic stellate cells promote immunotolerance following orthotopic liver transplantation in rats via induction of T cell apoptosis and regulation of Th2/Th3-like cell cytokine production. Experimental and Therapeutic Medicine. 2013;**5**(1):165-169

[21] Wu SL, Pan CE. Tolerance and chimerism and allogeneic bone marrow/stem cell transplantation in liver transplantation. World Journal of Gastroenterology. 2013;**19**(36):5981-5987. DOI: 10.3748/wjg.v19.i36.5981

[22] Liu XQ, Hu ZQ, Pei YF, Tao R. Clinical operational tolerance in liver transplantation: State-of-the-art perspective and future prospects. Hepatobiliary & Pancreatic Diseases International. 2013;**12**(1):12-33

[23] Nakano T, Lai CY, Goto S, Hsu LW, Kawamoto S, Ono K, et al. Immunological and regenerative aspects of hepatic mast cells in liver allograft rejection and tolerance. PloS One. [serial on the Internet]. 2012;**7**(5):e37202. DOI: 10.1371/journal.pone.0037202

[24] Schildberg FA, Hegenbarth SI, Schumak B, Scholz K, Limmer A, Knolle PA. Liver sinusoidal endothelial cells veto CD8 T cell activation by antigen-presenting dendritic cells. European Journal of Immunology. 2008;**38**(4):957-967

[25] Bettelli E, Carrier Y, Gao W, Korn T, Strom TB, Oukka M, et al. Reciprocal developmental pathways for the generation of pathogenic effector TH17 and regulatory T cells. Nature. 2006;**441**(7090):235-238

[26] Marie JC, Liggitt D, Rudensky AY. Cellular mechanisms of fatal early-onset autoimmunity in mice with the T cell-specific targeting of transforming growth factor-beta receptor. Immunity. 2006;**25**(3):441-454

[27] Li MO, Wan YY, Flavell RA. T cell-produced transforming growth factor-beta1 controls T cell tolerance and regulates Th1- and Th17-cell differentiation. Immunity. 2007;**26**(5):579-591

[28] Sanchez-Fueyo A, Strom TB. Immunologic basis of graft rejection

and tolerance following transplantation of liver or other solid organs. Gastroenterology. 2011;**140**(1):51-64. DOI: 10.1053/j.gastro.2010.10.059

[29] Charles R, Chou HS, Wang L, Fung JJ, Lu L, Qian S. Human hepatic stellate cells inhibit T-cell response through B7-H1 pathway. Transplantation. 2013;**96**(1):17-24. DOI: 10.1097/TP.0b013e318294caae

[30] Takahashi Y. The role of growth hormone and insulin-like growth factor-I in the Liver. International Journal of Molecular Sciences. 2017;**18**(7):1447. DOI: 10.3390/ijms18071447

[31] Stefano JT, Correa-Giannella ML, Ribeiro CM, Alves VA, Massarollo PC, Machado MC, et al. Increased hepatic expression of insulin-like growth factor-I receptor in chronic hepatitis C. World Journal of Gastroenterology. 2006;**12**(24):3821-3828

[32] Nishizawa H, Iguchi G, Fukuoka H, Takahashi M, Suda K, Bando H, et al. IGF-I induces senescence of hepatic stellate cells and limits fibrosis in a p53-dependent manner. Scientific Reports. 2016;**6**:34605. DOI: 10.1038/srep34605

[33] Choi JS, Park YJ, Kim SW. Three-dimensional differentiated human mesenchymal stem cells exhibit robust antifibrotic potential and ameliorates mouse liver fibrosis. Cell Transplantation. 2021;**30**:963689720987525. DOI: 10.1177/0963689720987525

[34] Lee SJ, Kim KH, Park KK. Mechanisms of fibrogenesis in liver cirrhosis: The molecular aspects of epithelial-mesenchymal transition. World Journal of Hepatology. 2014;**6**(4):207-216. DOI: 10.4254/wjh.v6.i4.207

[35] Krizhanovsky V, Yon M, Dickins RA, Hearn S, Simon J, Miething C, et al. Senescence of activated stellate cells limits liver fibrosis. Cell. 2008;**134**(4):657-667. DOI: 10.1016/j.cell.2008.06.049

[36] Dichtel LE, Cordoba-Chacon J, Kineman RD. Growth hormone and insulin-like growth factor I regulation of nonalcoholic fatty liver disease. The Journal of Clinical Endocrinology and Metabolism. 2022;**107**(7):1812-1824. DOI: 1210/clinem/dgac088

[37] Liu Z, Cordoba-Chacon J, Kineman RD, Cronstein BN, Muzumdar R, Gong Z, et al. Growth hormone control of hepatic lipid metabolism. Diabetes. 2016;**65**(12):3598-3609. DOI: 10.2337/db16-0649

[38] Cao LH, Lu FM, Lu XJ, Zhu LY. Study on the relationship between insulin growth factor 1 and liver fibrosis in patients with chronic hepatitis C with type 2 diabetes mellitus. Journal of Cellular Biochemistry. 2018;**119**(11):9513-9518. DOI: 10.1002/jcb.27267

[39] Gariani K, Toso C, Philippe J, Orci LA. Effects of liver transplantation on endocrine function: A systematic review. Liver International. 2016;**10**(10):13158. DOI: 10.1111/liv.13158 Epub 2016 Jun 23

[40] Chishima S, Kogiso T, Matsushita N, Hashimoto E, Tokushige K. The relationship between the growth hormone/insulin-like growth factor system and the histological features of nonalcoholic fatty liver disease. Internal Medicine. 2017;**56**(5):473-480. DOI: 10.2169/internalmedicine.56.7626

[41] Bonefeld K, Moller S. Insulin-like growth factor-I and the liver. Liver

International. 2011;**31**(7):911-919. DOI: 10.1111/j.1478-3231.2010.02428.x

[42] Castro GR, Coelho JC, Parolin MB, Matias JE, de Freitas AC. Insulin-like growth factor I correlates with MELD and returns to normal level after liver transplantation. Annals of Transplantation. 2013;**18**:57-62. DOI: 10.12659/AOT.883819

[43] Sobrevals L, Rodriguez C, Romero-Trevejo JL, Gondi G, Monreal I, Paneda A, et al. Insulin-like growth factor I gene transfer to cirrhotic liver induces fibrolysis and reduces fibrogenesis leading to cirrhosis reversion in rats. Hepatology. 2010;**51**(3):912-921. DOI: 10.1002/hep.23412

[44] de la Garza RG, Morales-Garza LA, Martin-Estal I, Castilla-Cortazar I. Insulin-like growth factor-1 deficiency and cirrhosis establishment. Journal of Clinical Medical Research. 2017;**9**(4):233-247. DOI: 10.14740/jocmr2761w

[45] Kawai M, Harada N, Takeyama H, Okajima K. Neutrophil elastase contributes to the development of ischemia/reperfusion-induced liver injury by decreasing the production of insulin-like growth factor-I in rats. Translational Research. 2010;**155**(6):294-304. DOI: 10.1016/j.trsl.2010.02.003

[46] Meng F, Zhang Z, Chen C, Liu Y, Yuan D, Hei Z, et al. PI3K/AKT activation attenuates acute kidney injury following liver transplantation by inducing FoxO3a nuclear export and deacetylation. Life Sciences. 2021;**272**(119119):26. DOI: 10.1016/j.lfs.2021.119119

[47] Conchillo M, de Knegt RJ, Payeras M, Quiroga J, Sangro B, Herrero JI, et al. Insulin-like growth factor I (IGF-I) replacement therapy increases albumin concentration in liver cirrhosis: Results of a pilot randomized controlled clinical trial. Journal of Hepatology. 2005;**43**(4):630-636. DOI: 10.1016/j.jhep.2005.03.025

[48] Bartke A. Growth hormone and aging: Updated review. World Journal of Men's Health. 2019;**37**(1):19-30. DOI: 10.5534/wjmh.180018

[49] Abdel-Wahab R, Hassan MM, George B, Carmagnani Pestana R, Xiao L, Lacin S, et al. Impact of integrating insulin-like growth factor 1 levels into model for end-stage liver disease score for survival prediction in hepatocellular carcinoma patients. Oncology. 2020;**98**(12):836-846. DOI: 10.1159/000502482

[50] Gautier S, Monakhov A, Tsiroulnikova O, Mironkov B, Voskanov M, Dzhanbekov T, et al. Time is of the essence: A single-center experience of hepatic arterial supply impairment management in pediatric liver transplant recipients. Pediatric Transplantation. 2021;**25**(3):12. DOI: 10.1111/petr.13934

[51] Cunningham EC, Sharland AF, Bishop GA. Liver transplant tolerance and its application to the clinic: Can we exploit the high dose effect? Clinical & Developmental Immunology. 2013;**419692**(10):6. DOI: 1155/2013/419692

[52] Aini W, Miyagawa-Hayashino A, Ozeki M, Tsuruyama T, Tamaki K, Uemoto S, et al. Frequent hepatocyte chimerism in long-term human liver allografts independent of graft outcome. Transplant Immunology. 2013;**28**(2-3):100-105. DOI: 10.1016/j.trim.2012.12.002

[53] Mogul DB, Luo X, Bowring MG, Chow EK, Massie AB, Schwarz KB, et al. Fifteen-year trends in pediatric liver transplants: Split, whole deceased, and living donor grafts. The Journal

of Pediatrics. 2018;**196**:148-153. DOI: 10.1016/j.jpeds.2017.11.015

[54] Monakhov A, Gautier S, Tsiroulnikova O, Semash K, Latypov R, Dzhanbekov T, et al. Living donor left lateral sectionectomy: Should the procedure still be performed open? Journal of Liver Transplantation. 2021;**1**:100001. DOI: 10.1016/j.liver.2020.100001

[55] WHO. Child Growth Standards based on length/height, weight and age. Acta Paediatrica. Supplement. 2006;**450**:76-85. DOI: 10.1111/j.1651-2227.2006.tb02378.x

[56] McDiarmid SV, Anand R, Lindblad AS. Development of a pediatric end-stage liver disease score to predict poor outcome in children awaiting liver transplantation. Transplantation. 2002;**74**(2):173-181. DOI: 10.1097/00007890-200207270-00006

[57] Freeman RB Jr, Wiesner RH, Roberts JP, McDiarmid S, Dykstra DM, Merion RM. Improving liver allocation: MELD and PELD. American Journal of Transplantation. 2004;**9**:114-131

[58] Kuemmerle JF. Insulin-like growth factors in the gastrointestinal tract and liver. Endocrinology and Metabolism Clinics of North America. 2012;**41**(2):409-423. DOI: 10.1016/j.ecl.2012.04.018

[59] Conchillo M, Prieto J, Quiroga J. Insulin-like growth factor I (IGF-I) and liver cirrhosis. Revista Española de Enfermedades Digestivas. 2007;**99**(3):156-164

[60] Shevchenko OP, Tsirulnikova OM, Tsirulnikova IE, Kurabekova RM, Olefirenko GA, Stepanova OI, et al. Dynamics of insulin-like growth factor-1 (IGF-1) in children after AB0-incompatible liver transplantation. Vestnik transplantologii i iskusstvennykh organov. 2014;**XVI**(2):46-51. DOI: 10.15825/1995-1191-2014-2-46-51

[61] Shevchenko OP, Tsirulnikova OM, Bougrov AV, Lourie YE, Tsirulnikova IE, Mnatsakanyan DS, et al. Insulin-like growth factor-1 after orthotopic liver transplantation in children with congenital and hereditary diseases of hepatobiliary system. Vestnik transplantologii i iskusstvennych organov. 2012;**XIV**(1):50-54. DOI: 10.15825/1995-1191-2012-1-50-54

[62] Roelfsema F, Veldhuis JD. Growth hormone dynamics in healthy adults are related to age and sex and strongly dependent on body mass index. Neuroendocrinology. 2016;**103**(3-4):335-344. DOI: 10.1159/000438904

[63] Jagodzinski C, Mueller S, Kluck R, Froede K, Pavičić L, Gellermann J, et al. Growth hormone treatment in the pre-transplant period is associated with superior outcome after pediatric kidney transplantation. Pediatric Nephrology. 2022;**37**(4):859-869. DOI: 10.1007/s00467-021-05222-5

Chapter 5

Carbohydrate Metabolism in Growth Hormone Therapy for Children

Mithal Dhbea and Abdilya Alabdaly

Abstract

Growth hormone deficiency is one of the most common causes of short stature in response to growth hormone therapy, and deficiency occurs as a result of either a decrease in the pituitary hormones themselves, which is called hypopituitarism, or a deficiency of hypothalamus. Stunting is a condition that worries many parents because if the body's growth hormone deficiency is not diagnosed and the appropriate treatment is not taken early, which leads to a high body mass index, that is, weight gain after puberty, high fat and future risks. For type 2 diabetes, diabetes, insulin-dependent heart disease, atherosclerosis and other diseases, the incidence of this deficiency in European societies is about 4000/1 births which is comparable with the proportion in Iraq but increasing the incidence. This study was conducted to investigate the effect of growth hormone therapy on carbohydrate metabolism where growth hormone affects the shape and function of the developing body and apart from these functions such as stimulating growth, it has distinctive effects on the metabolism and energy, and plays a role in regulating blood sugar levels.

Keywords: growth hormone, carbohydrate, metabolism, hypothalamus, pituitary

1. Introduction

Growth hormone (GH) has effect in carbohydrate metabolism including the stimulation of glucose production from the liver, in particular when it is abundant. The hormone also increases the extent to which beta cells in the Langerhans pancreatic cells can secrete insulin in response to glucose alert [1]. It was found that the daily injection of dogs with this hormone leads to diabetes with total atrophy in the islands of Langerhans as a result of the demoralization of the continued alertness of the islands of Langerhans, which leads to exhaustion due to excessive secretion of the hormone insulin caused by injection pituitary extracts [2]. Because the growth of growth hormone in dogs and cats accompanies the incidence of diabetes, it is not surprising to note the emergence of diabetes in all patients with Acromegaly. In general [3], insulin and growth hormone can be considered as complementary agents, each regulating the supply of energy to the tissues of the body where Insulin becomes active immediately after eating (a time when glucose as the most important source of energy in the body is forming) while growth hormone becomes active in hunger or fasting [4]. Insulin and Insulin-like growth factor-I (IGF-I) are the two hormones that control glucose metabolism; hence, IGF-I is the second factor after insulin that has an antihypertensive effect [5]. The growth element IGF-I, a polypeptide hormone with 70 amino acids and a molecular weight of 7649 Da, is a member of a big family that is highly similar to one another and is very close to the proportion of insulin [6]. It indirectly affects the growth of many cells and tissues in the body. IGF-amino I's acid sequence was discovered to be 48% similar to that of human proinsulin hormone, often known as "insulin-like," which is mostly released from the liver, according to Humble and Rinderknecht's research from 1978. Laron [7], where the body's nutritional situation is the main secretion. Additional hormones include the most potent one, growth hormone (GH), as well as insulin, thyroxin, and anabolic steroids [8]. Growth hormone is secreted under joint and coordinated control of three hypothalamus hormones: growth hormone-releasing hormone (GHRH). Somatostatin and Ghrelin (GHRH), GHRH and gerlin are positive regulators that stimulate growth hormone release [9]. Somatostatin (SST) is a negative regulator that inhibits the secretion of the hormone. Another factor is insulin-like growth factor (IGF-I), which affects the thalamus and pituitary, reducing growth hormone secretion [10]. Hypoglycemia is a potent stimulant for GH secretion. At the same time, hyperglycemia prevents the secretion of hormone in most cases [11]. Some report that a hypoglycemia level of 50% of normal level promotes hormone secretion. In secretion [12]. Causing Fasting for a long time increase the secretion of the hormone and after eating begins three stages of secretion of the hormone as the secretion of the hormone insulin is predominantly in the first stage while in the second phase less secretion of insulin and increases the secretion of growth hormone and in the third stage is the secretion of growth hormone is predominantly, also Physical exertion causes increased secretion of the hormone and there is a direct relationship between the effort exerted and the amount of the hormone secreted [13]. Some neurotransmitters such as fear, anxiety and noise may activate hormone secretion. It is also observed that hormonal factors influence GH secretion Thyroxin extracted from the thyroid gland, increases the amount of growth hormone releasing factor (GHRF) extracted from the hypothalamus, which increases the secretion of GH from the anterior lobe of the pituitary anterior pituitary gland [14]. The incidence of growth hormone deficiency in European societies is about 1/4000 births, which is similar to the rate in Iraq [15]. However, the increase in diabetes and other endocrine diseases in the context of wars and economic siege, and the increase in children with growth hormone deficiency of the reviewers of the unit of hormone The growth in the women's and children's education hospital in Ramadi as a result of war, displacement, lack of health care, interruption of treatment and diagnosis exacerbated this problem. Due to the lack of studies at the level of Al-Anbar province on the condition of lack of growth hormone in children with short stature therefore, this study was conducted in order to investigate the effect of therapeutic hormone growth hormone on the metabolism of carbohydrates, fats and bones in a group of children with hormone deficiency hormone (growth hormone deficiency only) and for 6 months of treatment, where growth hormone affects the form and function of the developing body and regardless of These functions as a stimulus to growth have specific effects on metabolism and energy, affect the fat cells to reduce the amount of stored fat, promote protein synthesis in cells and play a role in regulating blood sugar levels [16]. GH is necessary for natural growth and cell proliferation and regeneration in humans and some other organisms.

Apart from these functions as growth stimuli, it has distinct effects on metabolism and energy affects fat cells to reduce the amount of stored fat, promotes protein synthesis in cells and plays a role in regulating Blood sugar levels [17]. The insufficient secretion of growth hormone in the early periods of the life of the individual causes the weakness of the skeleton and the incidence of dwarfism, while excessive secretion of GH before the completion of growth or before the closure of rectangular bones causes the increase of the growth of these bones to a large extent, which is expressed by the disease Gigantism [18]. If the excessive increase in excretion happens in adulthood, the so- called Acromegaly will develop, which is characterized by the swelling of the bones of the head and jaw, hands and feet [3]. In fact, the increase in the secretion of GH after the epiphyseal cartilage is locked (calcified) causes the previous condition, which also the continued growth of bones without Epiphyses (under the influence of the increase in the hormone) contributes to, making the body organs ultimately inconsistent. In general, there are several centers for the formation of the human and animal skeleton, including the aforementioned Epiphyseal cartilage centers. These centers become active under the influence of GH where their cells divide and increase in fish while some of these cells are constantly transformed into bones [19].

Growth hormone is not directly affected by the insulin-like growth factor or somatomedin, as the growth hormone stimulates its secretion in peripheral tissues, especially the liver [2]. GH is believed to perform its functions on the cartilage (multiplying cartilage cells) indirectly by regulating the level of Somatomedins in blood plasma, some of which are known as insulin- like growth factors. These somatomedins therefore have anabolic effects in the cartilage in terms of activating the transfer of amino acids and stimulating the process of protein synthesis, DNA and RNA and increasing absorption of sulfate [7].

2. Growth hormone

Growth hormone (GH), also known as somatotropin, is a hormone that regulates healthy body development and growth. This process is made possible by accelerating the creation of protein in muscle cells and the release of energy from the breakdown of fat. A number of hormones are crucial to human development (GH), also known as somatotropin [20]. It is a 191 amino acid protein that the anterior pituitary gland secretes to regulate healthy growth and development, as depicted in the **Figure 1**. Human growth is divided into two phases. One at conception, the other at puberty. GH is important in these two stages [22]. Every day of a normal person's life,

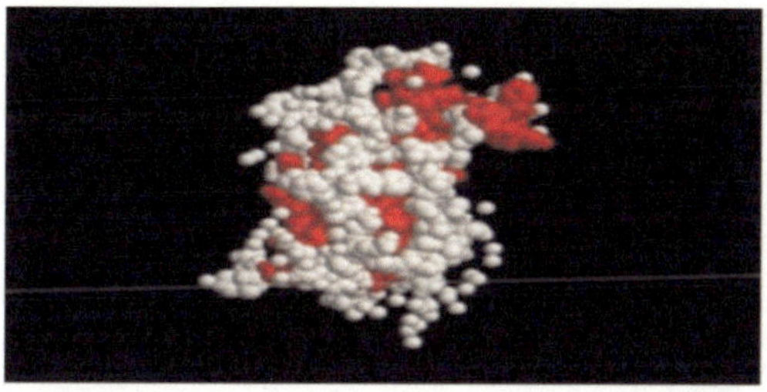

Figure 1.
Growth hormone [21].

amounts of GH are detectable. Nevertheless, because eating and activity have an impact on GH levels, they change throughout the day. There are GH-responsive receptors on the cells and tissues of the body [13]. The metabolic effects of GH are in muscle, liver, and fat cells, which are crucial to their function, but the visible effect of GH is on linear skeletal development [23]. There are two types of GH in humans. It's unclear how the two vary functionally, though. Although they are both products of the same gene, one of them lacks the amino acids between positions 32 to 46 [24] in Thongh.

Somatotrophs in the anterior pituitary gland create growth hormone (GH) under the guidance of hormonal signals in the hypothalamus [25]. GH is regulated by two hypothalamic hormones. GHRH (growth hormone-releasing hormone) and GHIH (growth hormone-inhibiting hormone) are the two (GHIH). This procedure' mechanism may be fully explained. As a result, when blood glucose levels decrease. The secretion of reserve GH is induced by GHRH. GHRH release is inhibited as blood glucose levels rise [26]. The effect is comparable when blood protein levels rise. This hypothalamic feedback loop causes GH levels to change during the course of the day. 1 to 3 ng/ML of normal plasma GH levels. The hormones insulin, glucagon, and adrenaline also control the availability of plasma glucose and amino acids for growth [27]. GH is primarily released during night. At 10 o'clock, midnight, and 2 a.m., GH release peaks. The rationale behind this time of day is that other hormones, such as the somatomedins, IGH-1 and IGH-2, mediate the majority of GH effects. As a result, GH's effects are more uniformly distributed throughout the day [25]. Due to their vital involvement in the production of GH and other hormones, a number of hormonal disorders can result in excessive or reduced development. However, a pituitary tumor frequently results in abnormal growth. Underproduction of GH, a deficiency in IGH-1, or a problem with the target tissue's reaction to either of these growth hormones can all contribute to dwarfism (extremely short stature). Gigantism and acromegaly, both of which are characterized by a very enormous stature, can result from an excessive production of GH or IGH-1, or from an exaggerated response to these hormones [28]. Early childhood overproduction of GH, which can result in gigantism, can generate skeletal heights of up to 8 feet (2.5 meters) or more [3]. When GH is overproduced after the onset of puberty, acromegaly develops. The body's long bone's epiphyseal plates are unable to close in this situation. In addition, they continue to respond to GH-stimulated further growth. An enlarged nose, throat, tongue, hands, feet, and cranium are features of this condition [28].

2.1 Growth hormone chemistry

A single chain polypeptide hormone of 191 amino acids and a molecular weight of 22,000 Daltons, human growth hormone is depicted in **Figure 2**. Four helices in the structure are required for a proper interaction with the GH receptor [31]. Growth hormone, chorionic somatomammotropin (placental lactogen), and prolactin have similar sequences.

Only the placenta secretes the 22-KD GH variant (hGH-V), which differs from pituitary GH by 13 amino acids [32]. Approximately 75% of the pituitary's typical GH secretion is in the mature 22-KD form. The loss of amino acids 32 to 46 caused by the second codon's alternative splicing produces in a 20 KD variant, which normally makes up 5 to 10% of pituitary GH.

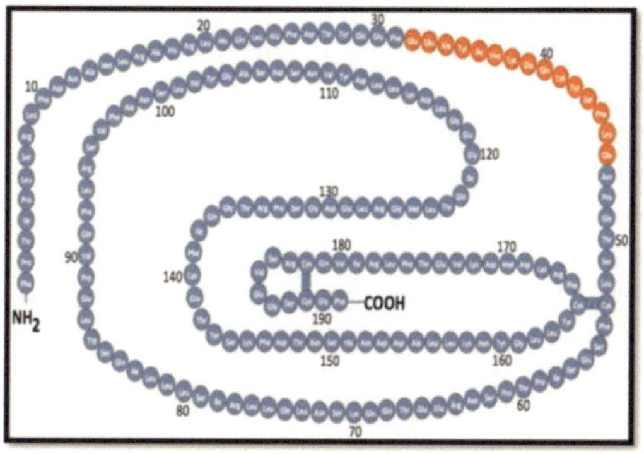

Figure 2.
Structure of growth hormone [29]. Pituitary GH is also found in deaminated and N-acetylated forms, as well as different GH oligomers [30].

2.2 Secretion of GH

Growth hormone (GH) is secreted into the blood by the somatotrope cells in the anterior pituitary gland as shown in the **Table 1**, more than any other pituitary hormone in significant quantities. These cells' growth and GH synthesis are both accelerated by the transcription factor PIT-1. GH shortage is the result of the anterior pituitary gland being destroyed and these cells failing to grow [33]. The main regulators of GH secretion by somatotropes are peptides produced by neurosecretory nuclei of the hypothalamus into the portal venous circulation surrounding the pituitary. Arcuate nucleus growth hormone-releasing hormone (GHRH) and Somatostatin from the periventricular nucleus suppresses GH secretion while ghrelin promotes it.

Hypothalamic Arcuate nucleus ← periventricular nucleus	**Growth hormone**
GHRH	↑
Somatostatine	↓
Aging and obesity	↑↓
Circulating GH, IGF-1	↓
Physiological: Sleep Exercise Hypoglycemia Dietary protein Estradiol	↑
Dietary carbohydrate Glucocorticoids	↓

Table 1.
Secretion of growth hormone.

Negative feedback from circulating GH and IGF-1 concentrations also affects GH secretion [20]. Even though the ratio of this stimulating to inhibitory peptide regulates GH release, various physiological stimulators and inhibitors of GH secretion can alter this ratio. Among other things, sleep, exercise, hypoglycemia, dietary protein, and estrogen are stimulators of GH secretion. Glucocorticoids and dietary carbohydrates are GH secretion inhibitors [22]. Throughout life, GH secretion follows a different rhythm. Early childhood has the highest basal levels. During the pubertal development spurt, the amplitude and frequency of peaks are at their highest. A healthy youngster or adolescent experiences eight peaks on average every day. Adults typically have five peaks. Throughout adulthood, basal levels, as well as the frequency and amplitude of peaks, decrease [34].

2.3 Function of GH

The actions of growth hormone on the body's tissues are referred to as anabolic effects (building up). GH, like the majority of other protein hormones, works by binding to a specific receptor on the surface of cells. The most well-known effect of GH activity is childhood height growth. There are at least two processes that seem to promote it [20]. Direct stimulation from GH causes cartilage chondrocytes to divide and multiply. These are the main cells of children's growing long bones (arms, legs, and digits), known as epiphyses [22]. In addition, GH promotes the synthesis of the proinsulin-like hormone insulin-like growth factor 1 (IGF 1, originally known as somatomedin C). The liver, which is the main location of IGF-1 synthesis, is a significant target organ of GH in this process. According to the table, IGF-1 exerts growth-stimulating effects on a range of tissues as shown in the **Table 2**. Target tissues produce more IGF-1, demonstrating that it is both an endocrine and an autocrine/paracrine hormone [22]. Although height growth is the most well-known consequence of GH, it also performs a variety of other metabolic tasks. GH improves bone mineralization, bone strength, and calcium absorption. Additionally, it builds muscle. Additionally, it stimulates the body's many organ systems to expand and produce more protein, creating a "positive nitrogen balance". GH boosts immunological function [20]. GH's part in maintaining fuel homeostasis. A counterproductive impact of GH was a reduction in the liver's absorption of glucose. It also helps keep pancreatic islets healthy and functioning. GH tends to encourage lipolysis, which causes a small decrease in adipose tissue (body fat) and an increase in the blood's levels of free fatty acids and glycerol [33].

Growth hormone effects		
Epiphysis (chondrocyte): 1. Linear growth before puberty. 2. Increase bone density and mineralization.	Liver: Secretion of IGF-1 (somatomedicine-c) Autocrine-paracrine hormone.	Target tissues: 1. Increase muscle mass. 2. Positive nitrogen balance. 3. Stimulate immune system. 4. Decrease hepatic glucose uptake. 5. lipolysis: Increase fatty acid in blood. Increase glycerol in blood.

Table 2.
Effect of growth hormone.

2.4 Genetic factors

2.4.1 Factors controlling growth

The aspects of growth are influenced by genetic mechanism, which is polygenic in nature. The first mechanism is the regulation of the rate cellular multiplication, affecting the overall size. The second is the control of the pace of maturation [35]. Both x and y chromosomes contain growth regulating genes. The special influence of X-chromosome is referred to in growth studies in females, asaril, girls develop 2 years earlier than boys [36]. Genetic factors such as body size are also known to be influenced by race. Black children stature and pace of osseous maturation during the childhood years.

2.4.2 Tissue-specific factors

The mechanism, by which control cellular multiplication, differentiation, and organ development during embryonic period of fetal life, remain a mystery. The concept that classic hormones directly regulate growth and maturation has been replaced by the hypothesis, that growth is controlled by variety of growth factors. Those factors production is stimulated by the classic hormones, these considered as mediator hormones because they mediate some of the biological actions of classic hormone [37].

2.4.3 Nutrition

Retardation of growth occurs, if malnutrition is severe, and in such cases there is always the additional possibility of chronic diseases such as anemia, congenital heart diseases etc. Improved nutrition and freedom from chronic disease are factors which have contributed to taller stature and earlier puberty of European and American prevalent, while poor linear growth, and delay adolescence are common among the children in the countries where malnutrition is prevalent. Chronic malnutrition through the childhood and early adult years, of prospective mothers, places their infants in jeopardy physically and intellectually [35].

2.5 Hormone AL regulation

In healthy children, statured growth is controlled by the action of circulating hormones. On skeletal system, GH and thyroid hormone (T4) are the major determinants of growth rate during the childhood [38]. They are working synergistically. T4 is predominantly required for skeletal maturation and GH for linear growth. While growth sprout and skeletal maturation of adolescent are primarily depending on gonad steroid in conjunction with GH and T4. Insulin and glucocorticiods influence carbohydrate, fat and protein metabolism, which provides sources of energy needed for growth and exerts a permissive influence on anabolic action of GH [39].

2.6 Growth hormone actions

Growth hormone (GH) has a wide range of biological effects, many of which are carried out directly by the GH-R and indirectly by IGF-1. The majority, if

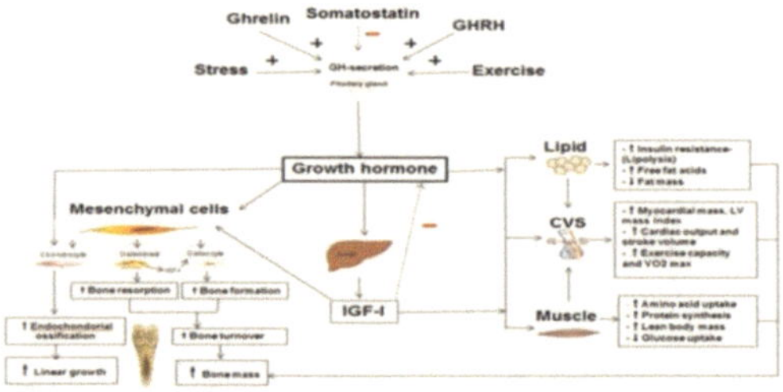

Figure 3.
The GH/IGF-1 axis and its effects on bone, muscle, and body metabolism. +, stimulation: −, inhibition: GHRH, Growth hormone-releasing hormone: cardiovascular system; VO2 Max: maximum oxygen consumption; CVS: cardiovascular system.

not all, tissues are affected by GH/IGF-1, which has no specific organ as a target as shown in the **Figure 3** [40].

2.6.1 Muscles growth and metabolism

By increasing protein synthesis and, presumably, limiting protein breakdown by up-regulating the production of Lipoprotein Lipase (LPL), both GH and IGF-1 regulate how quickly muscles break down protein [41, 42]. IGF-1 appears to control the size of human myotubes by promoting protein synthesis, preventing protein breakdown, and causing the reserve cells needed for maximum growth to fuse. Additionally, it has been found that GH causes muscular growth, which is most likely mediated by naturally occurring IGF-1's autocrine and paracrine effects [34]. Additionally, GH has been demonstrated to cause lipid buildup in the muscles, which has been well reported in people with excessive GH (acromegaly), causing skeletal muscle to switch from using glucose to lipids as a substrate [43]. Additionally, there is insufficient evidence—mostly from hypothetical animal studies—to support the idea that growth hormone (GH) may influence the composition of skeletal muscle fibers and cause a switch from glycolytic fast-twitch fibers (type II) to oxidative slow-twitch fibers (type I), which may indicate how GH affects muscle power and strength [44].

2.6.2 Protein metabolism

Growth hormone (GH) promotes protein synthesis while suppressing proteolysis either directly or through IGF endocrine and paracrines processes, which has an anabolic influence on protein metabolism [45, 46]. The majority of research point to GH's modest anabolic effects, which may include boosting body wide protein production while reducing protein breakdown, as well as reducing muscle-specific amino acid oxidation and degradation [1].

2.6.3 Glucose metabolism

Growth hormone (GH) is necessary to maintain the metabolism and homeostasis of glucose. While some of these actions are directly induced by IGF-1, others are mostly mediated through its effects that resemble those of insulin (in contrast to those of GH) [1]. The GH lipolytic impact appears to be the most important component of GH anti-insulin actions, preventing insulin-stimulated glucose absorption through the oxidation of FFA and consequent inhibition of glycolytic enzymes [47] Additionally, due to its lipolytic properties, GH produced anti-insulin effects that improved hepatic and peripheral insulin sensitivity, increased hepatic glucose production, decreased carbohydrate oxidation, and decreased insulin resistance [48]. Additional mechanisms for the metabolic effects of GH through downregulation of insulin signaling have been proposed, including increased expression of the p85 regulatory subunit of PI3K activity in adipose tissue and enhanced suppressors of cytokine signaling (namely SOCS 1 and 3). [49].

Factors which increase GH release.

1. The hypothalamic hormone growth hormone releasing hormone (GHRH) which is necessary for GH synthesis and stimulates GH release [50].

2. A second stimulatory system, parallel to that involving the GHRH receptor is activated that binds the GH secretagogue receptor to induce GHRH and GH [51]. Ghrelin is produced in the arcuate nucleus of the hypothalamus and in much greater quantity by the stomach, but also in the heart, lung and adipose tissue. Fasting increases gastric ghrelin gene expression Administration of ghrelin stimulates food intake and raises plasma GH concentration [52].

3. Sleep: The nighttime is when GH secretion peaks, particularly at the start of the first slow-wave sleep. But is connected to poor GH secretion [53].

4. Exercise: Exercise and physical stress include trauma with hypovolemic shock and sepsis increased GH levels [54].

5. Hypoglycemia: Chronic malnutrition and prolonged fasting increases the number and amplitude of GH secretion and enhanced GHRH release [51].

6. Protein intake: High protein meals and intravenous single amino acid (arginine leucine) stimulate growth hormone secretion [50].

2.7 Factors that inhibit GH release

Somatotropin release-inhibitory factor (SRIF) also known as somatostatin which inhibits.

GH release [37].

2.8 Obesity

Obesity is characterized by significantly increased GH synthesis, as evidenced by a nearly pronounced increase in the frequency of GH secretary bursts and half-life duration [55].

2.9 Growth hormone physiology and regulation

Growth hormone (GH, somatotropin) is secreted by somatotrophs which make up about 50 percent of anterior pituitary cells [56]. During these peaks, the plasma concentration of GH may range from 5 to 35 ng/ml or higher. Peaks may last for five to 30 minutes before resuming baseline levels. A hour or so after the start of sleep, these GH peaks reach their highest and most consistent heights [57]. There are many factors which modulate the production of growth hormone, including stress, nutrition, sleep as stimulator, and free fatty acids which are inhibitor of GH secretion. Several molecular isoforms of GH circulate in the plasma. Much of the growth hormone in the circulation is bound to a protein (growth hormone binding protein GHBP) which is derived from the growth hormone receptors [15]. Peptides released by neurosecretory nuclei of the hypothalamus (Growth hormone releasing hormone and somatostatin) into the portal venous blood surrounding the pituitary are the major controllers of GH secretion by the somatotrophs [31].

3. Insulin like growth factor-1 (IGF-1)

Insulin like growth factor-1 (IGF-1), and do Mains see polypeptide targets hormone for GH. The 70 amino acids IGF-1 protein consists of 4 domain as shown in the **Figure 4**. IGF-1 mediates most of the growth-promoting actions [59]. It is synthesized in the liver and extra hepatic tissue (principally muscle, bone, and kidney). It is a 7.7 kDa single-chain polypeptide that is also found in the pituitary gland [18, 60]. IGF-1 has an A and B chain that are joined by disulfide bonds, just like insulin does. The 12 amino acids in the C-peptide region. IGF-1's ability to attach to the insulin receptor (with low affinity) is demonstrated by its comparable structure to insulin [61]. IGF-1 provides negative feedback regulation of GH production and secretion by acting at the endocrine and paracrine level. The liver, high-affinity binding proteins like IGF-binding protein 3 (IGFBP3), and transport are where about 80% of IGF1 is produced. It controls IGF1 cell-surface receptor availability to mediate IGF-1 peptide activity as well [62].

Based on the person's age, circulating IGF-1 levels change. As a result of the decreased GH levels, there is initially an uptick from birth to puberty, which is then followed by a steady decline with age [63] IGFBP1–6, a group of six IGF-binding proteins, control the bioavailability and half-life of circulating IGF-1. Each IGFBP can bind to IGF-1 with a high degree of affinity, and it is controlled by a number of particular IGFBP proteases [64]. IGF-1 is a potent inhibitor of apoptosis and has a significant impact on cell proliferation and differentiation [65]. The liver is where IGF-1 is primarily produced [66]. Age and nutrition are just two of the variables that affect serum IGF-1 levels. However, GH is the main factor controlling IGF-1 release into the bloodstream and synthesis in the liver [67].

Figure 4.
Structure of IGF-1 [58].

3.1 Insulin-like growth factor-1(IGF-1) production and circulation

Insulin-like growth factor-1 is produced primarily by the liver. Then it is carried in blood bound to a carrier protein, which prolongs its half-life. Its level is therefore more constant than that of growth hormone [68]. Human growth hormone stays in the blood stream only for a short time, (few minutes). Every organ and system of the body is affected by growth hormone, either directly or indirectly, through the mediator IGF-1 [69]. Similar to that, it encourages young children's bone growth. The majority of organs and tissues experience growth, including the brain, which is also impacted. Most of the advantages and effects of human growth hormone are caused directly by IGF-1. IGF-1 has a ten-fold greater potency than HGH [70]. With a half-life of 21 to 40 hours. In addition, it is tightly bound to its binding protein in plasma. GH stimulates IGF-1 not only secretion but also the production of IGF binding protein [71]. IGF-1 mediates the anabolic growth promoting effects of GH in the muscle [72].

4. Carbohydrate metabolism

4.1 Growth hormone effect and treatment on the metabolism of carbohydrates

Growth hormone (GH) therapy blocks insulin's effects on peripheral tissues such skeletal muscle, the liver, and adipose tissue, increasing glucose synthesis from these tissues and reducing adipose tissue glucose uptake [73]. Following GH injection, insulin synthesis is enhanced to counteract the elevated blood glucose. Chronic exposure to high FFA may have a direct harmful effect on beta cells due to GH-induced visceral adipose tissue lipolysis and the subsequent increase in circulating FFA [74]. Because IGF-1 mimics insulin in the liver and skeletal muscle, increasing its levels following GH injection may improve insulin resistance and glucose balance [75]. The effects of GH treatment on the glucose metabolism of the pediatric population have only been the subject of a modest number of studies. The majority of these studies have shown that increased insulin resistance, as shown by elevated fasting insulin and homeostasis model assessment of insulin resistance levels, was seen in GH-deficient children and adolescents during GH therapy, but their fasting/postprandial glucose and HbA1c levels remained within in normal range [76].

In individuals with GH deficiency, GH injection has been linked to a reduction in visceral adiposity and an improvement in cardio-metabolic dysfunction, according to a number of human investigations. While taking GH, certain studies have raised concerns about increased insulin resistance and impaired fasting glucose, particularly in obese and older patients. Studies on children and teenagers have suggested that GH injection may result in short-term treatment-induced insulin resistance, although its long-term effects have not yet been extensively analyzed [77]. It is advisable to monitor any potential detrimental effects on glucose metabolism both before and after GH delivery because international cohort studies indicate that GH therapy may increase the prevalence of type 2 diabetes mellitus in children and teens with predisposed risk factors. The long-term effects of GH therapy on cardiovascular outcomes in GH-deficient children with or without GH continuing following end of skeletal development require extensive longitudinal cohort studies [78].

5. Glycosylated hemoglobin (HbA1c)

Hemoglobin that has been glycosylated (also known as hemoglobin A1c, or HbA1c) is one type of hemoglobin (Hb) that is primarily evaluated to determine the average blood glucose level over a long period of time. This protein is created by hemoglobin's regular exposure to plasma glucose in a gradual, nonenzymatic glycation reaction [79]. Hemoglobin that has been irreversibly glycosylated at either or both of the amino-terminal valines (Val) of the beta-polypeptide chains is referred to as glycosylated hemoglobin [80]. The most widely used and widely approved test for assessing the glycemic management of people with diabetes is HbA1c. Red blood cells (RBCs) retain the glycated form of hemoglobin (Hb) for the duration of the RBC's existence [81]. In clinical settings, HbA1c is employed to precisely reflect glycemic management over the previous (120 days). For individuals without diabetes, the optimal HbA1c test range is often less than 7%; for those with diabetes, the usual range is between 4 and 6%. [82]. The test can be taken at any time of day and requires no special preparation, such as fasting. These characteristics have made it the best test for assessing diabetics' glycemic control. The American Diabetes Association (ADA) now advises using HbA1c as a diagnostic test for diabetes as there has been much interest in doing so in recent years.

As a method for identifying those with a high risk of getting diabetes [83]. Furthermore, nephropathy and retinopathy in diabetes mellitus have been connected to hemoglobin glycosylation. Since its introduction in clinical settings in the 1980s, HbA1c has been a standard test for determining average plasma glucose over the preceding 8 to 12 weeks.

The relationship between HbA1c percent and plasma glucose levels is well recognized; the higher blood glucose levels, the higher the level of HbA1c percent, which is useful for monitoring and management; also, as HbA1c percent rises, the risk of complications from diabetes tends to rise. HbA1c% is a helpful indicator of how well plasma glucose level has been controlled recently and can be used to track the effects of exercise, diet, and medication therapy on plasma glucose in diabetic patients because HbA1c levels are unaffected by daily variations in plasma glucose levels and instead reflect average glucose levels over the prior 6 to 8 weeks [84]. The best tool available to doctors to assess the overall management of diabetes is HbA1c% as a consequence [85].

5.1 Insulin

A hormone produced in the pancreas is insulin (an organ located behind the stomach). Islets, or cell clusters, are found in the pancreas. Insulin is produced by beta cells within the islets and released into the circulation. In metabolism, insulin is important [86]. It is a key anabolic hormone that is secreted in response to increased amino acid blood and glucose followed by ingestion of a meals. Similar to many hormones, insulin works by attaching to certain receptors on the body's cells, including those in the fat, muscle, and liver cells [87]. Insulin's primary function is to promote the absorption of glucose. **Figure 5** shows the two polypeptide chains that make up the protein insulin: A (21 amino acid residues) and B (30 amino acid residues). Disulfide bridges connect chains A and B. An intrachain disulfide bridge connecting residues 6 and 11 [89] is also present in the A-chain.

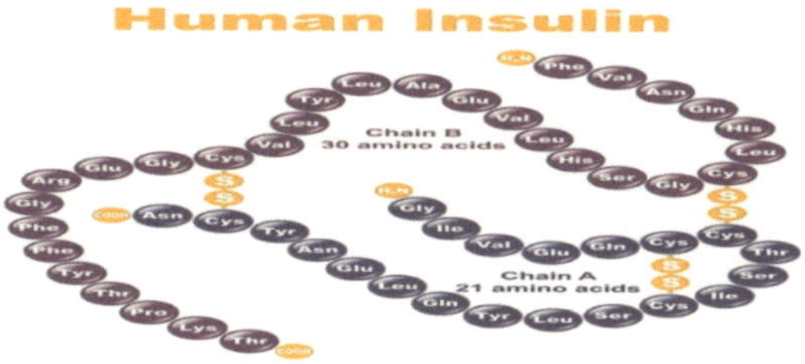

Figure 5.
Insulin structure [88].

5.2 Insulin resistance (IR)

Cells with insulin resistance (IR) do not react to the hormone's typical effects, which is a physiological condition [90, 91]. High blood sugar results from the body's cells becoming resistant to the insulin it generates, making them unable to utilize it properly [92]. A subsequent increase in insulin synthesis by beta cells in the pancreas furthers the rise in blood insulin levels [93].

Author details

Mithal Dhbea[1] and Abdilya Alabdaly[2]*

1 College of Medicine, University of Alanbar, Al-Anbar, Iraq

2 Collage of Medicine, University of Jordan, Jordan

*Address all correspondence to: abdily.riyadh@yahoo.com

References

[1] Rudman D. Growth hormone, body composition, and aging. Journal of the American Geriatrics Society. 1985; **33**(11):800-807

[2] Bustamante JJ, Gonzalez L, Carroll CA, Weintraub ST, Aguilar RM, Muñoz J, et al. O-glycosylated 24 kDa human growth hormone has a mucin-like biantennary disialylated tetrasaccharide attached at Thr-60. Proteomics. 2009;**9**(13):3474-3488

[3] King MW. Structure and Function of Hormones: Growth Hormone [Dissertation]. Indiana: Indiana State University; 2006

[4] Chellakooty M, Vangsgaard K, Larsen T, Scheike T, Falck-Larsen J, Legarth J, et al. A longitudinal study of intrauterine growth and the placental growth hormone (GH)- insulin-like growth factor I axis in maternal circulation: Association between placental GH and fetal growth. The Journal of Clinical Endocrinology and Metabolism. 2004;**89**(1):384-391

[5] Tetlow LJ, Clayton PE. Tests and normal values in pediatric endocrinology. Clinical Pediatric Endocrinology. 2005:523-564

[6] Juul A. Serum levels of insulin-like growth factor I and its binding proteins in health and disease. Growth Hormone & IGF Research. 2003;**13**(4): 113-170

[7] Laron Z. Laron syndrome: From description to therapy. Endocrinologist. 1993;**3**(1):21-28

[8] Martini FH, Nath JL, Bartholomew EF. Fundamentals of Anatomy and Physiology. New Jersey: Englewood Cliffs; 2009. p. 7632

[9] Freedman RJ, Malkovska V, LeRoith D, Collins MT. Hodgkin lymphoma in temporal association with growth hormone replacement. Endocrine Journal. 2005;**52**(5):571-575

[10] Ahmid MAE. Bone Health and Body Composition of Children and Adolescents with Growth Hormone Deficiency. Glasgow, Scotland: University of Glasgow; 2017

[11] Velloso CP. Regulation of muscle mass by growth hormone and IGF-I. British Journal of Pharmacology. 2008; **154**(3):557-568

[12] Chikani V, Ho KK. Action of GH on skeletal muscle function: Molecular and metabolic mechanisms. Journal of Molecular Endocrinology. 2014;**52**(1): R107-R123

[13] Freda PU, Shen W, Heymsfield SB, Reyes-Vidal CM, Geer EB, Bruce JN, et al. Lower visceral and subcutaneous but higher intermuscular adipose tissue depots in patients with growth hormone and insulin-like growth factor I excess due to acromegaly. The Journal of Clinical Endocrinology and Metabolism. 2008;**93**(6):2334-2343

[14] Sotiropoulos A, Ohanna M, Kedzia C, Menon RK, Kopchick JJ, Kelly PA, et al. Growth hormone promotes skeletal muscle cell fusion independent of insulin-like growth factor 1 up-regulation. Proceedings of the National Academy of Sciences. 2006; **103**(19):7315-7320

[15] Mauras N, O'Brien KO, Welch S, Rini A, Helgeson K, Vieira NE, et al. Insulin-like growth factor I and growth hormone (GH) treatment in GH-deficient humans: Differential effects on

protein, glucose, lipid, and calcium metabolism. The Journal of Clinical Endocrinology and Metabolism. 2000; **85**(4):1686-1694

[16] Mauras N, Haymond MW. Are the metabolic effects of GH and IGF-I separable? Growth Hormone & IGF Research. 2005;**15**(1):19-27

[17] Cronin MJ, Thorner MO. Basic studies with growth hormone-releasing factor. Endocrinology. 1989;**1**:183-191

[18] Hannah-Shmouni F, Trivellin G, Stratakis CA. Genetics of gigantism and acromegaly. Growth Hormone & IGF Research. 2016;**30**:37-41

[19] Cooke NE, Ray J, Watson MA, Estes PA, Kuo BA, Liebhaber SA. Human growth hormone gene and the highly homologous growth hormone variant gene display different splicing patterns. The Journal of Clinical Investigation. 1988;**82**(1):270-275

[20] Møller N, Jørgensen JOL. Effects of growth hormone on glucose, lipid, and protein metabolism in human subjects. Endocrine Reviews. 2009; **30**(2):152-177

[21] Am N, Jurayyan A, Sci BJ, Res T, Ah R, Khalifah A, et al. Childhood hypopituitarism: A three decades of experience from a major teaching Hospital in Central Region (Riyadh) Saudi. Arabia. 2018;**5**(4):8435-8441

[22] Jørgensen JOL, Krag M, Jessen N, Nørrelund H, Vestergaard ET, Møller N, et al. Growth hormone and glucose homeostasis. Hormone Research in Pædiatrics. 2004;**62**(Suppl. 3):51-55

[23] Piatti PM, Monti LD, Caumo A, Conti M, Magni F, Galli-Kienle M, et al. Mediation of the hepatic effects of growth hormone by its lipolytic activity. The Journal of Clinical Endocrinology and Metabolism. 1999;**84**(5):1658-1663

[24] Vijayakumar A, Novosyadlyy R, Wu Y, Yakar S, LeRoith D. Biological effects of growth hormone on carbohydrate and lipid metabolism. Growth Hormone & IGF Research. 2010;**20**(1):1-7

[25] Larsen R, Kronenberg H. Normal and aberrant growth. In: Williams Textbook of Endocrinology. 3rd ed. Philadelphia: Saunders; 2003. pp. 1038-1046

[26] Wren AM, Small CJ, Ward HL, Murphy KG, Dakin CL, Taheri S, et al. The novel hypothalamic peptide ghrelin stimulates food intake and growth hormone secretion. Endocrinology. 2000;**141**(11):4325-4328

[27] Tillmann V, Thalange NKS, Foster PJ, Gill MS, Price DA, Clayton PE. The relationship between stature, growth, and short-term changes in height and weight in normal prepubertal children. Pediatric Research. 1998;**44**(6): 882

[28] Holl RW, Hartman ML, Veldhuis JD, Taylor WM, Thorner MO. Thirty-second sampling of plasma growth hormone in man: Correlation with sleep stages. The Journal of Clinical Endocrinology and Metabolism. 1991;**72**(4):854-861

[29] Shariat N, Holladay CD, Cleary RK, Phillips JA III, Patton JG. Isolated growth hormone deficiency type II caused by a point mutation that alters both splice site strength and splicing enhancer function. Clinical Genetics. 2008;**74**(6):539-545

[30] Bona G, Paracchini R, Giordano M, Momigliano-Richiardi P. Genetic defects in GH synthesis and secretion. European Journal of Endocrinology. 2004;**151** (Suppl 1):S3-S9

[31] Argente J, Caballo N, Barrios V, Munoz MT, Pozo J, Chowen JA, et al. Multiple endocrine abnormalities of the growth hormone and insulin-like growth factor axis in patients with anorexia nervosa: Effect of short-and long-term weight recuperation. The Journal of Clinical Endocrinology and Metabolism. 1997;**82**(7):2084-2092

[32] Gardner CJ, Irwin A, Daousi C, MacFarlane I, Joseph F, Bell J, et al. Non alcoholic fatty liver disease, growth hormone deficiency and the effects of growth hormone replacement: A LIVERpool magnetic resonance spectroscopy study. European Journal of Endocrinology. 2012:EJE-12

[33] Acromegaly MS. The New England Journal of Medicine. 2006;**355**(24):2558-2573

[34] Mirjalili SA, Hinton L, Richards S. Endocrine and reproductive physiology. In: Physiology for General Surgical Sciences Examination (GSSE). Berlin: Springer; 2019. pp. 1-36

[35] Laron Z. Insulin-like growth factor 1 (IGF-1): A growth hormone. Molecular Pathology. 2001;**54**(5):311

[36] Melmed S. Acromegaly pathogenesis and treatment. The Journal of Clinical Investigation. 2009;**119**(11):3189-3202

[37] Cattani L, Ferri A. The function of NADPH bound to catalase. Bollettino della Società Italiana di Biologia Sperimentale. 1994;**70**(4):75-82

[38] Christopoulos PF, Msaouel P, Koutsilieris M. The role of the insulin-like growth factor-1 system in breast cancer. Molecular Cancer. 2015;**14**(1):1

[39] Philippou A, Maridaki M, Pneumaticos S, Koutsilieris M. The complexity of the IGF1 gene splicing, posttranslational modification and bioactivity. Molecular Medicine. 2014;**20**(1):202-214

[40] Fürstenberger G, Senn H-J. Insulin-like growth factors and cancer. The Lancet Oncology. 2002;**3**(5):298-302

[41] Aleem E, Elshayeb A, Elhabachi N, Mansour AR, Gowily A, Hela A. Serum IGFBP-3 is a more effective predictor than IGF-1 and IGF-2 for the development of hepatocellular carcinoma in patients with chronic HCV infection. Oncology Letters. 2012;**3**(3):704-712

[42] Rosenfeld RG, Kofoed E, Little B, Woods K, Buckway C, Pratt K, et al. Growth hormone insensitivity resulting from post-GH receptor defects. Growth Hormone & IGF Research. 2004;**14**:35-38

[43] Kofoed EM, Hwa V, Little B, Woods KA, Buckway CK, Tsubaki J, et al. Growth hormone insensitivity associated with a STAT5b mutation. The New England Journal of Medicine. 2003;**349**(12):1139-1147

[44] van der Spoel E, Jansen SW, Akintola AA, Ballieux BE, Cobbaert CM, Slagboom PE, et al. Growth hormone secretion is diminished and tightly controlled in humans enriched for familial longevity. Aging Cell. 2016;**15**(6):1126-1131

[45] Reynolds CM, Perry JK, Vickers MH. Manipulation of the growth hormone-insulin- like growth factor (GH-IGF) axis: A treatment strategy to reverse the effects of early life developmental programming. International Journal of Molecular Sciences. 2017;**18**(8):1729

[46] Bock GR, Goode JA. Biology of IGF-1: Its Interaction with Insulin in Health

and Malignant States. Vol. 262. Boca Raton: John Wiley & Sons; 2005

[47] Derek le Roith, Walter Zumkeller RCB. Insulin–Like Growth Factors, 4th edition, Berlin: Springer; 2003. p. 121-124

[48] Francis Sorrel Green Span DG. Basic and Clinical Endocrinology 7th edition, Greenspans Basic and Clinical Endocrinology. New York, USA: MC Graw–Hill Prifessional; 2004. p. 114-116

[49] Schwarz J-M, Mulligan K, Lee J, Lo JC, Wen M, Noor MA, et al. Effects of recombinant human growth hormone on hepatic lipid and carbohydrate metabolism in HIV-infected patients with fat accumulation. The Journal of Clinical Endocrinology and Metabolism. 2002;**87**(2):942-945

[50] Kovacs P, Stumvoll M. Fatty acids and insulin resistance in muscle and liver. Best Practice & Research. Clinical Endocrinology & Metabolism. 2005; **19**(4):625-635

[51] LeRoith D, Yakar S. Mechanisms of disease: Metabolic effects of growth hormone and insulin-like growth factor 1. Nature Reviews. Endocrinology. 2007; **3**(3):302

[52] Ciresi A, Cicciò F, Amato MC, Giordano C. Revaluation of the clinical and metabolic behavior of children with isolated growth hormone deficiency during GH treatment according to newly proposed note 39 of the Italian Medicines Agency (AIFA). Journal of Endocrinological Investigation. 2015; **38**(12):1301-1307

[53] Cutfield WS, Wilton P, Bennmarker H, Albertsson-Wikland K, Chatelain P, Ranke MB, et al. Incidence of diabetes mellitus and impaired glucose tolerance in children and adolescents receiving growth-hormone treatment. Lancet. 2000;**355**(9204): 610-613

[54] Child CJ, Zimmermann AG, Scott RS, Cutler GB Jr, Battelino T, Blum WF. Prevalence and incidence of diabetes mellitus in GH-treated children and adolescents: Analysis from the GeNeSIS observational research program. The Journal of Clinical Endocrinology and Metabolism. 2011; **96**(6):E1025-E1034

[55] Gupta S, Jain U, Chauhan N. Laboratory diagnosis of HbA1c: A review. Journal of Nanomedicine Research. 2017;**5**(4):120

[56] Syed I. Glycated haemoglobin; past, present, and future are we ready for the change. Diabetes. 2011;**11**:12

[57] Jain U, Gupta S, Chauhan N. Detection of glycated hemoglobin with voltammetric sensing amplified by 3D-structured nanocomposites. International Journal of Biological Macromolecules. 2017;**101**:896-903

[58] Bode BW, Braithwaite SS, Steed RD, Davidson PC. Intravenous insulin infusion therapy: Indications, methods, and transition to subcutaneous insulin therapy. Endocrine Practice. 2004;**10** (Supplement 2):71-80

[59] Pani LN, Korenda L, Meigs JB, Driver C, Chamany S, Fox CS, et al. Effect of aging on A1C levels in individuals without diabetes: Evidence from the Framingham offspring study and the National Health and nutrition examination survey 2001–2004. Diabetes Care. 2008;**31**(10):1991-1996

[60] Malkani S, Mordes JP. Implications of using hemoglobin A1C for diagnosing diabetes mellitus. The American Journal of Medicine. 2011;**124**(5):395-401

[61] Nathan DM, Turgeon H, Regan S. Relationship between glycated haemoglobin levels and mean glucose levels over time. Diabetologia. 2007;**50** (11):2239-2244

[62] Sigal RJ, Kenny GP, Boulé NG, Wells GA, Prud'homme D, Fortier M, et al. Effects of aerobic training, resistance training, or both on glycemic control in type 2 diabetes: A randomized trial. Annals of Internal Medicine 2007;147 (6):357–369

[63] Al-Sharifi ZAR. El-Yassin HD. Normal and obese healthy subjects. Journal of the Faculty of Medicine Baghdad. 2008;**50**(1):383-390

[64] Aronoff SL, Berkowitz K, Shreiner B, Want L. Glucose metabolism and regulation: Beyond insulin and glucagon. Diabetes Spectrum: A Publication of the American Diabetes Association. 2004;**17**(3): 183-190

[65] Joshi SR, Parikh RM, Das AK. Insulin-history, biochemistry, physiology and pharmacology. Journal-Association of Physicians of India. 2007; **55**(L):19

[66] Cheng P-Y, Sy H-N, Wu S-L, Wang W-F, Chen Y-Y. Newly diagnosed type 2 diabetes and risk of dementia: A population-based 7-year follow-up study in Taiwan. Journal of Diabetes and its Complications. 2012;**26**(5): 382-387

[67] Dobbins RL, Szczepaniak LS, Myhill J, Tamura Y, Uchino H, Giacca A, et al. The composition of dietary fat directly influences glucose-stimulated insulin secretion in rats. Diabetes. 2002; **51**(6):1825-1833

[68] Chiu HK, Tsai EC, Juneja R, Stoever J, Brooks-Worrell B, Goel A, et al. Equivalent insulin resistance in latent autoimmune diabetes in adults (LADA) and type 2 diabetic patients. Diabetes Research and Clinical Practice. 2007;**77**(2):237-244

[69] Colao A, Di Somma C, Cuocolo A, Spinelli L, Tedesco N, Pivonello R, et al. Improved cardiovascular risk factors and cardiac performance after 12 months of growth hormone (GH) replacement in young adult patients with GH deficiency. The Journal of Clinical Endocrinology and Metabolism. 2001; **86**(5):1874-1881

[70] Murray RD, Wieringa GE, Lissett CA, Darzy KH, Smethurst LE, Shalet SM. Low- dose GH replacement improves the adverse lipid profile associated with the adult GH deficiency syndrome. Clinical Endocrinology. 2002; **56**(4):525-532

[71] Lanes R, Gunczler P, Lopez E, Esaa S, Villaroel O, Revel-Chion R. Cardiac mass and function, carotid artery intima-media thickness, and lipoprotein levels in growth hormone-deficient adolescents. The Journal of Clinical Endocrinology and Metabolism. 2001;**86**(3): 1061-1065

[72] Lanes R, Paoli M, Carrillo E, Villaroel O, Palacios A. Cardiovascular risk of young growth-hormone-deficient adolescents. Hormone Research in Pædiatrics. 2003; **60**(6):291-296

[73] Al-Shoumer KAS, Cox KH, Hughes CL, Richmond W, Johnston DG. Fasting and postprandial lipid abnormalities in hypopituitary women receiving conventional replacement therapy. The Journal of Clinical Endocrinology and Metabolism. 1997; **82**(8):2653-2659

[74] Lanes R, Paoli M, Carrillo E, Villaroel O, Palacios A. Peripheral inflammatory and fibrinolytic markers in adolescents with growth hormone deficiency: Relation to postprandial dyslipidemia. The Journal of Pediatrics. 2004;**145**(5):657-661

[75] Robbins RJ, REICHLIN S. Somatostatin biosynthesis by cerebral cortical cells in monolayer culture. Endocrinology. 1983;**113**(2): 574-581

[76] Georg P, Ludvik B. Lipids and diabetes. Journal of Clinical and Basic Cardiology. 2000;**3**(3):159-162

[77] Kumar. Lipid profile: Parametric test sounds simple but significant in diagnostic approach. Asian Journal of Medical and Biological Research. 2018; **4**(1):1-4

[78] Miller NE. HDL metabolism and its role in lipid transport. European Heart Journal. 1990;**11** (suppl_H):1-3

[79] Hu G. Total cholesterol and the risk of Parkinson's disease: A review for some new findings. Park Dis. 2010;**2010**: 1972-1979

[80] Bays HE, McKenney J, Maki KC, Doyle RT, Carter RN, Stein E. Effects of prescription omega-3-acid ethyl esters on non—High-density lipoprotein cholesterol when coadministered with escalating doses of atorvastatin. In: Mayo Clinic Proceedings. Elsevier; 2010. pp. 122-128

[81] Orozco-Beltran D, Gil-Guillen VF, Redon J, Martin-Moreno JM, Pallares-Carratala V, Navarro-Perez J, et al. Lipid profile, cardiovascular disease and mortality in a Mediterranean high-risk population: The ESCARVAL-RISK study. PLoS One. 2017;**12**(10):e0186196

[82] O'Keefe JH, Cordain L, Harris WH, Moe RM, Vogel R. Optimal low-density lipoprotein is 50 to 70 mg/dl: Lower is better and physiologically normal. Journal of the American College of Cardiology. 2004;**43**(11):2142-2146

[83] Hussain MM. A proposed model for the assembly of chylomicrons. Atherosclerosis. 2000;**148**(1):1-15

[84] (EFSA) EFSA. Calcium and bone growth-scientific substantiation of a health claim related to calcium and bone growth pursuant to article 14 of regulation (EC) No 1924/2006-scientific opinion of the panel on dietetic products, nutrition and allergies. EFSA Journal. 2008;**6**(10):826

[85] Anderson JJB, Garner SC, Klemmer PJ. Diet, Nutrients, and Bone Health. CRC Press; 2016

[86] Fazlini MF, Suriah AR, Zaitun Y, Kandiah M, Chee WSS, Chan YM, et al. Calcium supplementation amongst postmenopausal women: Effect on serum calcium, phosphorus and magnesium level. International Food Research Journal. 2013;**20**(1)

[87] Bootman MD, Collins TJ, Peppiatt CM, Prothero LS, MacKenzie L, De Smet P, et al. Calcium signalling— An overview. In: Seminars in cell & developmental biology. Elsevier; 2001. pp. 3-10

[88] Holick FM. Vitamin D: Evolutionary, physiological and health perspectives. Current Drug Targets. 2011;**12**(1):4-18

[89] Holick MF. Sunlight and vitamin D for bone health and prevention of autoimmune diseases, cancers, and cardiovascular disease. The American Journal of Clinical Nutrition. 2004;**80** (6):1678S-1688S

[90] Armas LAG, Hollis BW, Heaney RP. Vitamin D2 is much less effective than vitamin D3 in humans. The Journal of Clinical Endocrinology and Metabolism. 2004;**89**(11):5387-5391

[91] Houghton LA, Vieth R. The case against ergocalciferol (vitamin D2) as a vitamin supplement. The American Journal of Clinical Nutrition. 2006; **84**(4):694-697

[92] Dhaked RK, Alam SI, Dixit A, Singh L. Purification and characterization of thermo- labile alkaline phosphatase from an Antarctic psychrotolerant bacillus sp. P9. Enzyme and Microbial Technology. 2005;**36**(7): 855-861

[93] Corathers SD. The alkaline phosphatase level: Nuances of a familiar test. Pediatrics in Review. 2006;**27**:382

Chapter 6

Obesity: The Relationship between Growth Hormone and Exercises

Danúbia da Cunha de Sá-Caputo, Mario Bernardo-Filho, Redha Taiar and Técia Maria de Oliveira Maranhão

Abstract

Obesity is one of the main causes of death around the world. Moreover, considering the cardiometabolic risk (CMR), the relationship between obesity and CMR is well-established, and the location of adipose tissue (AT), particularly in the abdominal region, is considered an important predictor of metabolic dysfunction than total fat mass. Central obesity can be related to abdominal subcutaneous adipose tissue (SAT) and visceral adipose tissue (VAT). The relationship between SAT and CMR is not still clear, but the VAT has been considered a unique pathogenic fat depot. In this context, it is important to identify clinical interventions that could be used to improve the management of obesity. The aim of this chapter is to integrate knowledge about the relevance of exercises and/or growth hormone (GH) to the management of individuals with obesity. In conclusion, it appears that exercise-induced reductions in VAT are mediated by induced changes in GH levels. This could be due to the similar lipolytic effects of both GH and exercise on VAT and this relationship would benefit the role of exercise as an intervention against obesity. Preventing and understanding the development of obesity is therefore essential if it is wanted to curb the global epidemic and save social security several million costs concerning health problems.

Keywords: obesity, growth hormone, exercise, abdominal subcutaneous adipose tissue, visceral adipose tissue

1. Introduction

Obesity is one of the main causes of death around the world. There has been a significant global increase in obesity rate during the last decades. This problem represents a global phenomenon occurring in all parts of the world region except parts of sub-Saharan Asia and Africa [1]. Moreover, this complex and undesirable clinical condition is a high- risk factor for various non-communicable diseases, such as, type 2 diabetes, cardiovascular disease, metabolic syndrome (MetSy), chronic kidney disease, hyperlipidemia, hypertension, nonalcoholic fatty liver disease, obstructive sleep apnea, osteoarthritis, and certain types of cancers [2, 3].

There are several possible mechanisms leading to obesity. The main cause is the significantly more excess energy stored in fat cells than the energy the body needs. This set characterizes the obesity disease [2, 4]. Research of Sacks et al. [5] showed that the food sources and quality of nutrients matter more than their quantity in the diet contributes for weight imbalance. The pathogenesis of obesity involves regulation of calorie utilization, appetite and physical activity, but have complex interactions with availability of health-care systems, the role of socio-economic status, and underlying hereditary and environmental factors [6].

Considering the cardiometabolic risk (CMR), the relationship between obesity and CMR is well-established, and the location of adipose tissue (AT), particularly in the abdominal region, is considered an important predictor of metabolic dysfunction than total fat mass [7, 8]. In obesity, the central obesity (CO) is characterized by the excess accumulation of AT in the abdominal region. The CO is strongly and independently correlated with MetSy and is assessed through the measurement of the waist circumference (WC) [8, 9].

As it is indicated in **Figure 1**, CO can be related to abdominal subcutaneous adipose tissue (SAT) and visceral adipose tissue (VAT) [10]. The relationship between SAT and CMR is not still clear, but, the VAT has been considered a unique pathogenic fat depot.

It is suggested that VAT would be a metabolic organ that would regulate fat mass, and glucose and nutrient homeostasis. VAT is also an active endocrine organ that synthesizes and secretes numerous bioactive mediators, adipocytokines/adipokines, and other vasoactive substances [11, 12]. These adipocytokines/adipokines act on various organs with metabolic importance, such as liver and skeletal muscle. There is evidence of the roles of adipocytokines/adipokines in the regulation of metabolic disorders like diabetes, obesity and insulin resistance. VAT is associated strongly with CMR independent of overall body mass index (BMI) or total body adiposity [13–15].

The growth of the economy in the world, mechanized transport, urbanization, industrialization, an increasingly sedentary lifestyle, and a nutritional transition to processed foods with high-calorie diets have favored the increase of the prevalence of obesity. Moreover, the rising prevalence of childhood obesity suggests a burden of this disease in the healthcare systems in the future [16].

In this context, it would be important to identify clinical interventions that could be used to improve the management of individuals with obesity. Considering that the relationship between CO and impaired growth hormone (GH) secretion, it is still poorly understood [17, 18], but the relevance of the exercise, as an intervention to improve the GH secretion has been highlighted [19–21].

Exercise and diet modification can be considered as therapies for the management of obesity. Verheggen et al. [19] reported that while exercise is less effective than diet for body mass loss, exercise would promote superior reductions in VAT. Moreover, Berryman and List [20] reinforce that this finding may partly be explained by exercise-induced changes in lipolytic hormones, such as GH, during and after exercise, which seem to target VAT.

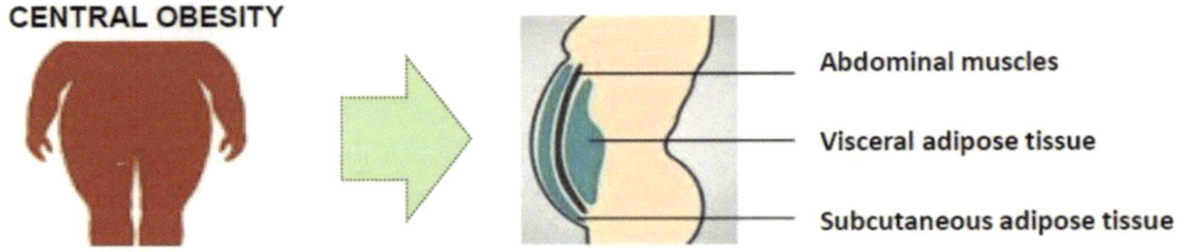

Figure 1.
Central obesity and subcutaneous and visceral adipose tissue.

Putting together these considerations, the aim of this chapter is to integrate knowledge about the relevance of exercises and/or GH to the management of individuals with obesity.

2. Growth hormone

GH is secreted at the anterior pituitary gland in a pulsatile manner and is primarily regulated by hypothalamic neuropeptides GH-releasing hormone and somatostatin, which stimulate and inhibit GH secretion, respectively [21]. It is known that resistance exercise (RE) is considered the most effective potent and physiological stimulus for GH release [22, 23], but little about how RE alters somatotroph content and function. Importantly, Rudman et al. [24] verified that when GH was administered to older healthy male, there were substantial decrease in AT mass, and significant increase in lean body mass. After that, studies have shown that GH therapy can improve VAT, circulating lipid levels, and insulin resistance in individuals with obesity and/or diabetes [15, 25]. In addition, it was observed the potential utility of GH therapy for the amelioration of age-related declines in metabolic function and body composition. GH is a potent anabolic hormone that affects multiple systems within the body and plays a significant role in lipid metabolism at various sites, such as liver, skeletal muscle, and AT [26]. Meanwhile, side effects of GH therapy such as an increased likelihood of soft tissue edema, joint pain, carpal tunnel syndrome, gynecomastia, hypertension, and diabetes have been reported [27, 28]. Considering these findings, exogenous GH therapy would became typically reserved for individuals with GH deficiencies resulting from hypothalamic/pituitary disease [29]. Despite this, there has since been increasing interest in identifying therapies, including lifestyle interventions, that increase physiologic GH release and action.

Considering the obesity, the somatotropic axis, that is a primary regulator of the metabolism, has particular relevance. The somatotropic axis consists of GH and insulin-like growth factors (IGF-I and IGF-II), and related to carrier proteins and receptors, which are further regulated by nutritional status and hormones such as ghrelin and insulin [30, 31]. During periods of fasting or stress, GH promotes the use of lipids as the primary fuel source in order to preserve carbohydrates and protein stores. In the liver, lipid uptake and production are increased through the phosphorylation of sterol regulatory element-binding proteins and by increased lipoprotein lipase (LPL) expression. In addition, GH also indirectly would increase the fatty acid oxidation and, additionally, would activate the adenosine monophosphate-activated protein kinase pathway [21].

GH is a powerful regulator of lipid metabolism, but its effect depends on the target. GH has lipogenic effects within the liver, however, the opposite occurs in AT, particularly VAT, where GH elicits lipolytic effects due to the suppression of LPL activity [17]. During exercise or fasting, GH stimulates the release of free fatty acids (FFAs) in the circulation to be delivered to various organs, including myocytes. In this case, FFAs may be repackaged as triglycerides or undergo β-oxidation in the mitochondria. While it is recognized that GH also elicits various effects on glucose and protein metabolism, exercise induced alterations in physiologic GH appear to primarily affect the AT lipolysis [32, 33].

Increased ectopic fat, such as VAT and intrahepatic triglyceride, contributes to insulin resistance and may affect the feedback control system of the somatotropic axis, resulting in a cascade of metabolic impairments [31].

3. Relationship between GH and exercise

Increases in GH secretion due to stress, such as fasting, or exercise contributes to lead increases in circulating FFAs [34]. Stokes et al. [35] showed that FFA levels may also regulate GH throughout a negative feedback control, as nicotinic acid-mediated suppression of lipolysis, and consequently reducing circulating FFAs, led to an important GH response. This finding may help further explain why individuals with obesity and reduced cardiorespiratory fitness (CRF), who on average have elevated levels of FFAs [36, 37].

Regular aerobic exercise enhances the ability of the body in the transportation and oxidization of FFAs during exercise [38]. This is also found in individuals with impaired fatty acid oxidation, such as those with obesity and diabetes [38–40], and it has been suggested that these improvements may be mediated through exercise-induced increases in mitochondrial and fatty acid transporter content, carnitine shuttle activity, and CRF [41, 42].

It is reported that low CRF may be a greater predictor of metabolic dysfunction than VAT [43], and as such, improving CRF has emerged as a therapeutic target for individuals with obesity-related disease. Moreover, acute exercise would temporarily increase the GH release, and such responses would be mediated by CRF [4]. Furthermore, although both aerobic and resistance exercise elicit a GH response, the relative contribution of aerobic exercise on GH response and action arguably is not totally understood. In this context, the evaluation of factors that contribute to aerobic exercise-induced GH response and how these changes influence VAT and cardiometabolic health more broadly have been studied [4].

GH promotes lipolysis within AT and increases mitochondrial oxidative capacity [44, 45] Although obesity decreases the exercise-induced GH response, CRF, which, at least, partly, is related to the muscle oxidative capacity, would be a more relevant determinant of exercise-induced GH secretion [4, 46].

Although the exercise and GH eliciting similar effects on AT and lipid metabolism, it is unclear whether exercise induced desirable effects in central adiposity are mediated by changes in physiologic GH response or if these effects in central adiposity and GH response are independent of exercise adherence [33, 47].

Acute aerobic exercise has been shown to increase GH levels, and these changes have been shown to be strongly associated with exercise intensity and volume, a function of exercise duration and frequency [48, 49]. It is suggested that exercise-induced GH responses may only be elicited at, or above, specific exercise volume and intensity parameters. Sasaki et al. [50] reported that with high-intensity interval training (HIIT) or moderate-intensity continuous training (MICT), the magnitude of GH response to exercise did not increase from pre-intervention measures in sedentary but otherwise healthy men for both interventions. Zhang et al. [51] reported, in a randomized controlled trial with young women with obesity, that when compared to energy-matched MICT, HIIT or supramaximal aerobic exercise led to greater VAT reduction but not greater changes from pre-intervention levels in serum GH measured immediately or 4 h after exercise. In fact, all groups showed elevated GH responses to exercise; however, only the higher-intensity interventions decreased VAT, suggesting that other factors likely contributed to these improvements. In addition, Calixto et al. [52] reported that the velocity of eccentric muscle action alters the acute responses following bench press exercises performed by resistance-trained men with a slow velocity leading to greater metabolic stress and GH response. Jørgensen et al. [53], previously, have pointed out that GH stimulates lipolysis and lipid oxidation during basal and fasting conditions and investigated whether GH also regulates substrate metabolism during exercise. The GH-deficient individuals were studied during exercise with and without GH administration as compared to untreated healthy subjects. It was verified that the GH predominantly stimulated the turnover of free fatty acids in the recovery phase after exercise. Then, it is possible to verify that aerobic and anaerobic exercises influence of GH secretion [51–53] and, consequently, these exercises could have positive effects on the CO.

Whole-body vibration exercise (WBVE), used in systemic vibratory therapy [54], is a type of physical exercise. In the WBVE, mechanical vibration (MV) generated in the vibrating platform is transmitted to the body of an individual that is in contact with this platform. Some publications have reported that the WBVE can alter the release of GH [55–58]. The potential physiological effects of WBVE on various organs/tissues would be also related to possible neuromuscular responses and the tonic vibratory reflex [59–61]. Moreover, the interaction of the MV with the mechanosensory system would be also involved in the effects described by the WBVE. The mechanosensors in the body cells, once stimulated by the MV, modulate the biological activity through specific signaling pathways, such as releasing hormones, like GH, and other substances (e.g., amino acids, proteins, lipids, ions) [54]. In consequence, several biological effects might be observed in several organs and tissues due to the WBVE [59, 61]. Furthermore, it is relevant to highlight that WBVE, depending on the parameters used in the protocols of the interventions and posture, can be considered aerobic or anaerobic exercises [62, 63].

The mechanisms driving the exercise-induced improvements in cardiometabolic outcomes and the relation with the GH is not fully clear yet. But, due to the various metabolic impacts of the exercise and GH, it is presented in **Figure 2** a possible relationship among GH and exercise, VAT and obesity.

Like GH, exercise exerts potent lipolytic effects, particularly on VAT [59]. Furthermore, exercise elicits improvements in ectopic fat in the absence of body

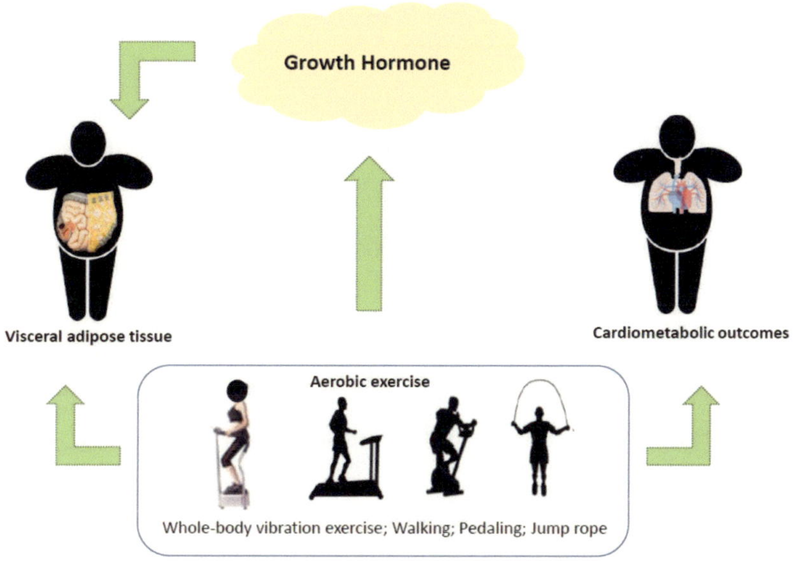

Figure 2.
The proposed relationship among growth hormone, physical exercises, adipose tissue, and cardiometabolic outcomes. Green lines represent positive effects from physical exercise to visceral adipose tissue.

mass loss [4]. As exercise induced GH response occurs in an intensity-dependent manner, and appears to be mediated by CRF, this may explain why HIIT can lead to similar improvements in WC and VAT than in higher-volume MICT despite requiring less time and expending less energy [60]. Moreover, considering the physiological adaptations and general health benefits of HIIT, individuals with obesity would be encouraged to perform such exercise.

4. Conclusion

In conclusion, it appears that exercise-induced reductions in VAT are mediated by induced changes in GH levels. This could be due to the similar lipolytic effects of both GH and exercise on VAT and this relationship would benefit the role of different types of exercises (aerobic and anaerobic exercises) as interventions against the obesity. Preventing and understanding the development of obesity is therefore essential, if it is wanted to curb the global epidemic and save social security several million costs concerning health problems.

Acknowledgements

The authors thank Conselho Nacional de Desenvolvimento Científico e Tecnológico (CNPq), Coordenação de Aperfeiçoamento de Pessoal de Nível Superior (CAPES), and Fundação de Amparo á Pesquisa do Estado do Rio de Janeiro (FAPERJ), Brazil, for the support.

Author details

Danúbia da Cunha de Sá-Caputo[1], Mario Bernardo-Filho[1*], Redha Taiar[2] and Técia Maria de Oliveira Maranhão[3]

1 Laboratory of Mechanical Vibrations and Integrative Practices, Roberto Alcantara Gomes Biology Institute, Piquet Carneiro University Polyclinic, State University of Rio de Janeiro, Rio de Janeiro, Brazil

2 Materials and Mechanical Engineering Research Group, University of Reims, Reims, France

3 Department of Obstetrics and Gynecology, Center of Health Sciences, Federal University of Rio Grande do Norte, Natal, Brazil

*Address all correspondence to: bernardofilhom@gmail.com

References

[1] Al Kibria GM. Prevalence and factors affecting underweight, overweight and obesity using Asian and World Health Organization Cutoffs among adults in Nepal: Analysis of the demographic and health survey 2016. Obesity Research & Clinical Practice. 2019;**13**(2):129-136. DOI: 10.1016/j.orcp.2019.01.006

[2] Swinburn BA, Sacks G, Hall KD, McPherson K, Finegood DT, Moodie ML, et al. The global obesity pandemic: Shaped by global drivers and local environments. Lancet. 2011;**378**(9793):804-814. DOI: 10.1016/S0140-6736(11)60813-1

[3] Frühbeck G, Toplak H, Woodward E, Yumuk V, Maislos M, Oppert JM. Obesity: The gateway to ill health—An EASO position statement on a rising public health, clinical, and scientific challenge in Europe. Obesity Facts. 2013;**6**:117-120. DOI: 10.1159/000350627

[4] Sabag A, Chang D, Johnson NA. Growth hormone as a potential mediator of aerobic exercise-induced reductions in visceral adipose tissue. Frontiers in Physiology. 2021;**26**(12):623570. DOI: 10.3389/fphys.2021.623570

[5] Sacks FM, Bray GA, Carey VJ, Smith SR, Ryan DH, Anton SD, et al. Comparison of weight-loss diets with different compositions of fat, protein, and carbohydrates. New England Journal of Medicine. 2009;**360**(9):859-873. DOI: 10.1056/NEJMoa0804748

[6] Lin X, Li H. Obesity: Epidemiology, pathophysiology,and therapeutics. Frontiers in Endocrinology. 2021;**12**:706978. DOI: 10.3389/fendo.2021.706978

[7] Kahn SE, Hull RL, Utzschneider KM. Mechanisms linking obesity to insulin resistance and type 2 diabetes. Nature. 2006;**444**:840-846. DOI: 10.1038/nature05482

[8] Chait A, den Hartigh LJ. Adipose tissue distribution, inflammation and its metabolic consequences, including diabetes and cardiovascular disease. Frontiers in Cardiovascular Medicine. 2020;**25**(7):22. DOI: 10.3389/fcvm.2020.00022

[9] Shen W, Punyanitya M, Chen J, Gallagher D, Albu J, Pi-Sunyer X, et al. Waist circumference correlates with metabolic syndrome indicators better than percentage fat. Obesity (Silver Spring). 2006;**2006**(14):727-736. DOI: 10.1038/oby.2006.83

[10] Snel M, Jonker JT, Schoones J, Lamb H, De Roos A, Pijl H, et al. Ectopic fat and insulin resistance: Pathophysiology and effect of diet and lifestyle interventions. International Journal of Endocrinology. 2012:983814. DOI: 10.1155/2012/983814

[11] Kanaya AM, Harris T, Goodpaster BH, Tylavsky F, Cummings SR. Adipocytokines attenuate the association between visceral adiposity and diabetes in older adults. Diabetes Care. 2004;**27**:1375-1380. DOI: 10.2337/diacare.27.6.1375

[12] Vasamsetti SB, Natarajan N, Sadaf S, Florentin J, Dutta P. Regulation of cardiovascular health and disease by visceral adipose tissue-derived metabolic hormones. Journal of Physiology. 2022. DOI: 10.1113/JP282728

[13] Fox CS, Massaro JM, Hoffmann U, Pou KM, Maurovich Horvat P,

Liu CY, et al. Abdominal visceral and subcutaneous adipose tissue compartments: Association with metabolic risk factors in the Framingham heart study. Circulation. 2007;**116**:39-48. DOI: 10.1161/CIRCULATIONAHA.106.675355

[14] Pak K, Lee SH, Lee JG, Seok JW, Kim IJ. Comparison of visceral fat measures with cardiometabolic risk factors in healthy adults. PLoS One. 2016;**11**:e0153031. DOI: 10.1371/journal.pone.0153031

[15] Saxton SN, Clark BJ, Withers SB, Eringa EC, Heagerty AM. Mechanistic links between obesity, diabetes, and blood pressure: Role of perivascular adipose tissue. Physiology Reviews. 2019;**99**(4):1701-1763. DOI: 10.1152/physrev.00034.2018

[16] Hruby A, Hu FB. The epidemiology of obesity: A big picture. PharmacoEconomics. 2015;**33**(7):673-689. DOI: 10.1007/s40273-014-0243-x

[17] Stanley TL, Grinspoon SK. Effects of growth hormone-releasing hormone on visceral fat, metabolic, and cardiovascular indices in human studies. Growth Hormones IGF Research. 2015;**25**:59-65. DOI: 10.1016/j.ghir.2014.12.005

[18] Lewitt MS. The role of the growth hormone/insulin like growth factor system in visceral adiposity. Biochemistry Insights. 2017;**10**:1178626417703995-1178626417703995. DOI: 10.1177/1178626417703995

[19] Verheggen RJ, Maessen MF, Green DJ, Hermus AR, Hopman MT, Thijssen DH. A systematic review and meta-analysis on the effects of exercise training versus hypocaloric diet: Distinct effects on body weight and visceral adipose tissue. Obesity Reviews. 2016;**17**:664-690. DOI: 10.1111/obr.12406

[20] Berryman DE, List EO. Growth hormone's effect on adipose tissue: Quality versus quantity. International Journal of Molecular Sciences. 2017;**18**:1621. DOI: 10.3390/ijms18081621

[21] Vijayakumar A, Yakar S, Leroith D. The intricate role of growth hormone in metabolism. Frontiers in Endocrinolology. (Lausanne). 2011;**2**:32. DOI: 10.3389/fendo.2011.00032

[22] Nicholls AR, Holt RI. Growth hormone and insulin-like growth factor-1. Frontiers in Hormone Research. 2016;**47**:101-114. DOI: 10.1159/000445173

[23] Fink JE, Schoenfeld BJ, Kikuchi N, Nakazato K. Acute and long-term responses to different rest intervals in low-load resistance training. International Journal of Sports Medicine. 2017;**38**(2):118-124. DOI: 10.1055/s-0042-119204

[24] Rudman D, Feller AG, Nagraj HS, Gergans GA, Lalitha PY, Goldberg AF, et al. Effects of human growth hormone in men over 60 years old. New England Journal of Medicine. 1990;**323**:1-6. DOI: 10.1056/NEJM199007053230101

[25] Nam SY, Kim KR, Cha BS, Song YD, Lim SK, Lee HC, et al. Low-dose growth hormone treatment combined with diet restriction decreases insulin resistance by reducing visceral fat and increasing muscle mass in obese type 2 diabetic patients. International Journal of Obesity and Related Metabolism Disorders. 2001;**25**:1101-1107. DOI: 10.1038/sj.ijo.0801636

[26] Dehkhoda F, Lee CMM, Medina J, Brooks AJ. The growth hormone receptor: Mechanism of receptor activation, cell signaling, and physiological aspects. Frontiers in

Endocrinology (Lausanne). 2018;**9**:35. DOI: 10.3389/fendo.2018.00035

[27] Liu H, Bravata DM, Olkin I, Nayak S, Roberts B, Garber AM, et al. Systematic review: The safety and efficacy of growth hormone in the healthy elderly. Annals of Internal Medicine. 2007;**2007**(146):104-115. DOI: 10.7326/0003-4819-146-2-200701160-00005

[28] Divall SA, Radovick S. Growth hormone and treatment controversy; long term safety of rGH. Current Pediatric Reports. 2013;**1**(2):128-132. DOI: 10.1007/s40124-013-0009-5

[29] Clemmons DR, Molitch M, Hoffman AR, Klibanski A, Strasburger CJ, Kleinberg DL, et al. Growth hormone should be used only for approved indications. The Journal of Clinical Endocrinology & Metabolism. 2014;**99**:409-411. DOI: 10.1210/jc.2013-4187

[30] Renaville R, Hammadi M, Portetelle D. Role of the somatotropic axis in the mammalian metabolism. Domestic Animal Endocrinology. 2002;**23**:351-360. DOI: 10.1016/S0739-7240(02)00170-4

[31] Savastano S, Di Somma C, Barrea L, Colao A. The complex relationship between obesity and the somatropic axis: The long and winding road. Growth Hormone IGF Research. 2014;**24**:221-226. DOI: 10.1016/j.ghir.2014.09.002

[32] Kanaley JA, Dall R, Møller N, Nielsen SC, Christiansen JS, Jensen MD, et al. Acute exposure to GH during exercise stimulates the turnover of free fatty acids in GH-deficient men. Journal of Applied Physiology. 1985;**2004**(96):747-753. DOI: 10.1152/japplphysiol.00711.2003

[33] Kopchick JJ, Berryman DE, Puri V, Lee KY, Jorgensen JOL. The effects of growth hormone on adipose tissue: Old observations, new mechanisms. Nature Reviews Endocrinology. 2020;**16**(3):135-146. DOI: 10.1038/s41574-019-0280-9

[34] Huang Z, Huang L, Waters MJ, Chen C. Insulin and growth hormone balance: Implications for obesity. Trends in Endocrinology & Metabolism. 2020;**31**:642-654. DOI: 10.1016/j.tem.2020.04.005

[35] Stokes KA, Tyler C, Gilbert KL. The growth hormone response to repeated bouts of sprint exercise with and without suppression of lipolysis in men. Journal of Applied Physiology. 1985;**2008**(104):724-728. DOI: 10.1152/japplphysiol.00534.2007

[36] Boden G. Obesity and free fatty acids. Endocrinology and Metabolism Clinics of North America. 2008;**37**:635-646. DOI: 10.1016/j.ecl.2008.06.007

[37] Manna P, Jain SK. Obesity, oxidative stress, adipose tissue dysfunction, and the associated health risks: Causes and therapeutic strategies. Metabolic Syndrome and Related Disorders. 2015;**13**(10):423-444. DOI: 10.1089/met.2015.0095

[38] Van Tienen FH, Praet SF, De Feyter HM, Van Den Broek NM, Lindsey PJ, Schoonderwoerd KG, et al. Physical activity is the key determinant of skeletal muscle mitochondrial function in type 2 diabetes. The Journal of Clinical Endocrinology & Metabolism. 2012;**97**:3261-3269. DOI: 10.1210/jc.2011-3454

[39] Ghanassia E, Brun JF, Fedou C, Raynaud E, Mercier J. Substrate oxidation during exercise: Type 2 diabetes is associated with a decrease in lipid oxidation and an earlier shift towards carbohydrate utilization. Diabetes & Metabolism. 2006;**32**:604-610. DOI: 10.1016/S1262-3636(07)70315-4

[40] Kolnes KJ, Petersen MH, Lien-Iversen T, Højlund K, Jensen J. Effect of exercise training on fat loss-energetic perspectives and the role of improved adipose tissue function and body fat distribution. Frontiers in Physiology. 2021;**12**:737709. DOI: 10.3389/fphys.2021.737709

[41] Melanson EL, Maclean PS, Hill JO. Exercise improves fat metabolism in muscle but does not increase 24-h fat oxidation. Exercise and Sport Sciences Reviews. 2009;**37**:93-101. DOI: 10.1097/JES.0b013e31819c2f0b

[42] Kujala UM, Vaara JP, Kainulainen H, Vasankari T, Vaara E, Kyröläinen H. Associations of aerobic fitness and maximal muscular strength with metabolites in young men. JAMA Network Open. 2019;**2**:e198265. DOI: 10.1001/jamanetworkopen.2019.8265

[43] Kim S, Kim JY, Lee DC, Lee HS, Lee JW, Jeon JY. Combined impact of cardiorespiratory fitness and visceral adiposity on metabolic syndrome in overweight and obese adults in Korea. PLoS One. 2014;**9**:e85742. DOI: 10.1371/journal.pone.0085742

[44] Short KR, Moller N, Bigelow ML, Coenen-Schimke J, Nair KS. Enhancement of muscle mitochondrial function by growth hormone. The Journal of Clinical Endocrinology & Metabolism. 2008;**93**:597-604. DOI: 10.1210/jc.2007-1814

[45] Takahashi Y. The role of growth hormone and insulin-like growth factor-I in the liver. International Journal of Molecular Sciences. 2017;**18**(7):1447. DOI: doi.org/10.3390/ijms18071447

[46] Thomas GA, Kraemer WJ, Comstock BA, Dunn-Lewis C, Maresh CM, Volek JS. Obesity, growth hormone and exercise. Sports Medicine. 2013;**43**(9):839-849. DOI: 10.1007/s40279-013-0064-7

[47] Thompson D, Karpe F, Lafontan M, Frayn K. Physical activity and exercise in the regulation of human adipose tissue physiology. Physiolology Reviews. 2012;**92**(1):157-191. DOI: 10.1152/physrev.00012.2011

[48] Veldhuis JD, Olson TP, Takahashi PY, Miles JM, Joyner MJ, Yang RJ, et al. Multipathway modulation of exercise and glucose stress effects upon GH secretion in healthy men. Metabolism. 2015;**64**(9):1022-1030. DOI: 10.1016/j.metabol.2015.05.008

[49] Atakan MM, Li Y, Koşar ŞN, Turnagöl HH, Yan X. Evidence-based effects of high-intensity interval training on exercise capacity and health: A review with historical perspective. International Journal of Environmental Research and Public Health. 2021;**18**:7201. DOI: doi.org/10.3390/ijerph18137201

[50] Sasaki H, Morishima T, Hasegawa Y, Mori A, Ijichi T, Kurihara T, et al. 4 weeks of high-intensity interval training does not alter the exercise induced growth hormone response in sedentary men. Springerplus. 2014;**3**:336. DOI: 10.1186/2193-1801-3-336

[51] Zhang H, Tong TK, Kong Z, Shi Q, Liu Y, Nie J. Exercise training-induced visceral fat loss in obese women: The role of training intensity and modality. Scandinavian Journal of Medicine & Science in Sports. 2021;**31**:30-43. DOI: 10.1111/sms.13803

[52] Calixto R, Verlengia R, Crisp A, Carvalho T, Crepaldi M, Pereira A, et al. Acute effects of movement velocity on blood lactate and growth hormone responses after eccentric bench press exercise in resistance-trained men.

Biology Sport. 2014;**31**(4):289-294. DOI: 10.5604/20831862.1127287

[53] Jørgensen JO, Krag M, Kanaley J, Møller J, Hansen TK, Møller N, et al. Exercise, hormones, and body temperature. Regulation and action of GH during exercise. Journal of Endocrinology Investigation. 2003;**26**(9):838-842. DOI: 10.1007/BF03345233

[54] Sá-Caputo DC, Seixas A, Taiar R, Bernardo-Filho M. Vibration therapy for health promotion. In: Bernardo-Filho M, Taiar R, Sá-Caputo DC, Seixas A, editors. Complementary Therapies. London: IntechOpen; 2022. DOI: 10.5772/intechopen.105024

[55] Paineiras-Domingos LL, Sá-Caputo DDC, Moreira-Marconi E, Morel DS, da Fontoura DC, Sousa-Gonçalves CR, et al. Can whole body vibration exercises affect growth hormone concentration? A systematic review. Growth Factors. 2017;**35**(4-5):189-200. DOI:10.1080/08977194.2017.1401619

[56] Rigamonti AE, Haenelt M, Bidlingmaier M, De Col A, Tamini S, Tringali G, et al. Obese adolescents exhibit a constant ratio of GH isoforms after whole body vibration and maximal voluntary contractions. BMC Endocrine Disorders. 2018;**18**(1):96. DOI: 10.1186/s12902-018-0323-6

[57] Weber-Rajek M, Mieszkowski J, Niespodziński B, Ciechanowska K. Whole-body vibration exercise in postmenopausal osteoporosis. Prz Menopauzalny. 2015;**1**:41-47. DOI: 10.5114/pm.2015.48679

[58] Giunta M, Rigamonti AE, Agosti F, Patrizi A, Compri E, Cardinale M, et al. Combination of external load and whole body vibration potentiates the GH-releasing effect of squatting in healthy females. Hormone and Metabolic Research. 2013;**45**(8):611-616. DOI: 10.1055/s-0033-1341464

[59] Rittweger J. Vibration as an exercise modality: How it may work, and what its potential might be. European Journal of Applied Physiology. 2010;**108**(5):877-904. DOI: 10.1007/s00421-009-1303-3

[60] Cochrane DJ. Vibration exercise: The potential benefits. International Journal of Sports Medicine. 2011;**32**(2):75-99. DOI: 10.1055/s-0030-1268010

[61] van Heuvelen MJG, Rittweger J, Judex S, Sañudo B, Seixas A, Fuermaier ABM, et al. Reporting guidelines for whole-body vibration studies in humans, animals and cell cultures: A consensus statement from an International Group of Experts. Biology (Basel). 2021;**10**(10):965. DOI: 10.3390/biology10100965

[62] Tsiloulis T, Watt MJ. Exercise and the regulation of adipose tissue metabolism. Progress in Molecular Biology and Translational Science. 2015;**135**:175-201. DOI: 10.1016/bs.pmbts.2015.06.016

[63] Keating SE, Hackett DA, Parker HM, O'connor HT, Gerofi JA, Sainsbury A, et al. Effect of aerobic exercise training dose on liver fat and visceral adiposity. Journal of Hepatolology. 2015;**63**:174-182. DOI: 10.1016/j.jhep.2015.02.022